AF650378

TEXTBOOK OF BIOCHEMISTRY

[According to Latest Syllabus of First Semester B.Sc. (Biochemistry) of Bangalore University]

By

Dr. Naveen Chandra

&

Dr. B.V. Somashekaraiah

ANMOL PUBLICATIONS PVT. LTD.

NEW DELHI - 110 002 (INDIA)

ANMOL PUBLICATIONS PVT. LTD.
[illegible] Ansari Road, Daryaganj
New Delhi-110 002
[illegible]
[illegible]

Textbook of B[illegible]
[illegible]

[illegible] 2005

[All rights reserved. No part of this publication may be reproduced, stored in a retrieval system or transmitted, in any form or by any means, electronic, mechanical, photocopying, recording or otherwise, without prior written permission of the publisher.]

PRINTED IN INDIA

Published by J.L. Kumar for Anmol Publications Pvt. Ltd., New Delhi - 110 002 and Printed at Mehra Offset Press, Delhi.

PREFACE

Chemistry is the subject flooded with lot of information for the past many decades. This is because the principles of chemistry is applicable to all walks of life. In many branch of sciences like Agricultural science, Veterinary science, Medical science, Pharmaceutical science, Microbiology, Molecular biology etc. Chemistry plays an important role in understanding the above branches of science.

There appears to be general agreement that courses in chemistry should be restructured in such a way as to present interdisciplinary principles of chemistry. This inter disciplinary approach is essential in presenting chemistry to beginners. The traditional structure of physical, organic, inorganic and analytical chemistry has origins in the main activities of chemists in the past, but continues to determine how chemistry is taught. Chemistry has developed into an applied science and people trained as chemists are making significant contributions in areas which overlap strongly with physics, biology as well as earth and material sciences.

The need for formulating proper chemistry curriculum suited to college students in this country was recognised by scientific community. Keeping all these facts in view, the UGC had proposed a nation wide uniform syllabus for undergraduate courses.

However, Bangalore university has restructured the syllabus of chemistry to be taught at undergraduate level. The new chemistry syllabus helps the undergraduate students to equip with fundamentals of chemistry and hence, it will lay the strong foundations for their higher goals.

In this text book, the information is given according to latest syllabus of I BSc., students of Biochemistry. The basic principles of chemistry which is necessary to understand Biochemical systems are presented in this book.

We have enjoyed and learnt lot of things while preparing this text book and we hope that it will be of great help for BSc., Biochemistry students and to our colleagues who are involved in teaching chemistry. Any suggestions or comments for the improvement of this book is gratefully accepted.

The authors

ACKNOWLEDGEMENTS

We would like to thank several friends and colleagues who have helped us in bringing out this text book.

We are greatful to Anmol publications Pvt. Ltd. for having accepted to publish this text book, which is written according to latest syllabus of Bangalore university. The Authors greatfully acknowledge Mr. Narayan Malkod, Sumukha Book House & Mr. Mahesh for having typed this manuscript.

1

MEASUREMENT

Scientists study and do research on the properties of matter and results are expressed based on certain definite measurements. Hence it is necessary for the students of science to understand how to express the results of the experiment they conduct in their laboratory. It may be noted that measurement is a fundamental means of communication. The measurement techniques are used by scientists for understanding the natural phenomenon and people of other sector use measurement technique for transacting business.

It becomes absolutely necessary that all the measured quantities and events of happenings must be widely defined that, no ambiguity is left to the exact meaning of any word phrase, other words it leads to endless confusion. We come across many different types of measurement and units in our daily life. Every measurement like distance, a weight, volume, an interval of time or any other measurements require two important things; (i) a number and (2) a unit or a standard. Quantity measured is compared with the standard to find out how many times the quantity is greater or less than the standard. The standard is generally called as the unit in which the quantity is measured. The measured number unit stands for how many time greater. A unit is a value, quantity or magnitude in terms of which other values or quantities or magnitudes are expressed. A unit is independent of physical definition. A standard should be such that it must be possible to define it unambiguously and it should be easily reproducible, it must be invariable with the time and space and also it should be possible to multiply or divide each one standard.

Fundamental quantities and units

The present day scientific field has achieved many milestone in theoretical concepts and experimental details of matter. In this age, the scientific and technical field have gathered a large number of varied body of data that have been accumulated. The number is increasing as research development increases. Hence we come across large number of unrelated scientific data and this would suggest that a large number units to be used to represent such huge scientific data. This would lead to an impossible task to handle different units meant for the quantities obtained by different area of scientific and technical fields. Therefore, this is not the case in practice, however, it is possible to express all the physical quantities in terms of three basic quantities. Also we find that more units are used in practice for the sake of convenience and not because it is necessary to do so.

The basic quantities for which separate distinct units which are essential are called fundamental quantities.

A fundamental quantity is a quantity which cannot be expressed in terms of any other physical quantity. The units in which other quantities are measured are called fundamental units.

Derived units

It may be noted here that in science, the quantities, except fundamental quantities are related in some manner to the fundamental quantities like mass, length and time. In reality all other quantities that can be expressed in terms of other quantities are known as derived quantities and the units in which these quantities are measured are called derived units. All those units which can be obtained by combining fundamental units like the units of area, volume, velocity, rate, acceleration, force, gravity etc are called the derived units. That means all these units have been derived from the fundamental units should become clear by considering an example as given below.

Area = Length x Breadth = length x length

Volume = Length x Breadth x height

$$\text{Velocity} = \frac{\text{Change in distance}}{\text{Time}}$$

$$= \frac{\text{Distance}}{\text{Time}} = \frac{\text{length}}{\text{Time}}$$

$$\text{Acceleration} = \frac{\text{Change in Velocity}}{\text{Time}}$$

$$= \frac{\text{Velocity}}{\text{Time}} = \frac{\text{Length}}{\text{Time x Time}} = \frac{\text{Length}}{(\text{Time})^2}$$

and Force = Mass × Acceleration

$$= \text{Mass} \times \frac{\text{Length}}{(\text{time})^2}$$

$$= \text{Force} = \frac{\text{Mass} \times \text{length}}{(\text{time})^2}$$

System of units

The following system have been used

1. The French system.
2. The CGS system
3. The British system
4. The M.K.S. system
5. The SI system.

CGS system: In this system, the units of length, mass and time are centimeter, gram and second respectively.

Centimeter: It is $\frac{1}{100}$ th part of a meter (m)

The gram It is $\frac{1}{1000}$ th part of a kilogram (kg)

Second It is defined as $\frac{1}{86400}$ th part of a mean solar day. One solarday is the time between two consecutive moons of the average time which elapses during a year between two successive passages of the sun across any one straight line drawn from pole to pole on the earth's surface.

F.P.S system. In this system, the unit of mass is pound (lb and the unit for time is the second. For length it is foot. The foot is $\frac{1}{3}$rd of the distance between two lines on platinum iridium Bar at a temperature of 62^0 F.

MKS System: In this system, the unit of mass is kilogram, the unit of length is meter and the unit of time is second.

System of international units (SI units)

This is the latest version of the metric system which is the most used all over the world. In this system, there are seven basic units, with the help of which we can derive units for all other possible quantities in science and engineering. In this system three more fundamental units are included and their additional units serve as link between science and engineering technological units. This system has an edge over the other three because it is comprehensive, coherent and rational. It is comprehensive in the sense that its six basic units cover all disciplines. The SI unit is found to cover all the areas of science where the measurement is required.

The Basic units used in SI system is represented in table 1

Physical quantity	*units*	*Abbreviation*
Length	Meter	m.
Mass	Kilogram	kg.
Time	Second	s
Electric current	Ampere	A
Luminous intensity	Candela	Cd
Temperature	Kelvin	K.
Amount of substance	Mole	mol.

The other units of measurements can be derived from these basic units. Like metric system of units, SI units are modified in decimal fashion by a series of prefixes.

Prefix	*Symbol*	*Meaning*
Deci	d	10^{-1}

Centi	c	10^{-2}
Milli	m	10^{-3}
Micro	m	10^{-6}
Nano	n	10^{-9}
Pico	p	10^{-12}

Mass and weight

Mass is measure of quantity of matter in an object. The terms mass and weight used interchangeably. However both mass and weight are different. Weight refers to the force that gravity exerts on an object. Mass of an object remains constant, therefore weight can vary.

Mass is expressed in terms of Kilogram (kg). But in chemistry the smaller unit gram is more convenient.

$$1Kg = 1000g = 1x10^3 \text{ g.}$$

Volume: Volume is length in meters (m) cubed therefore, the SI derived unit is the cubic meter (m^3).

But the chemists and Biochemists work with smallcr volumes such as cubic centimeters and cubic decimeter (dm^3)

$$1 \text{ cm}^3 = (1 \times 10^{-2} \text{ m})^3 = 1 \times 10^{-6} \text{ m}^3$$

$$1 \text{ dm}^3 = (1 \times 10^{-1} \text{ m})^3 = 1 \times 10^{-3} \text{ m}^3$$

Another common unit for volume is (L). L is the volume occupied by one cubic decimeter. One Liter of volume is equal to 1000 milliliters (mL) and one milliliter of volume is equal to one cubic centimeter.

$$1L = 1000ml$$

$$1mL = 1 \times 10^{-3} \text{ L}$$

$$IL = 1 \text{ dm}^3$$

$$1 \text{ dm}^3 = 1 \times 10^3 \text{ cm}^3$$

$$1 \text{ mL} = 1\text{cm}^3$$

Density

Density is the derived unit and it is the mass of the object divided by its volume.

$$\text{Density} = \frac{\text{Mass}}{\text{volume}}$$

$$d = \frac{M}{V}$$

The SI unit for density is kilogram per cubic meter (Kg/m^3). This unit is very large, therefore, smaller units like g/mL or g/L is used in chemistry and Bio chemistry experiments.

Temperature scale

There are three temperature scales used currently for the measurement of temperature. They are 1) Kelvin (K), 2) Degree Celsius (^{0}C) and Fahrenheit (^{0}F)

The SI unit of temperature is Kelvin. In 1848 Lord Kelvin identified that the temperature - 273.15^{0}c as the theoretically the lowest attainable temperature called absolute zero. With absolute zero as a starting point, he set up an absolute temperature scale now called kelvin temperature scale. On degree Celsius is equal in magnitude to one Kelvin (k). The only difference between the absolute temperature scale and the Celsius scale is that the zero position is shifted.

Absolute Zero = - 273.15^0 c

Freezing point of water 273.15K = 0^0 c.

Boiling point of water 373.15 = 100^0 c

Dimensional analysis

In Chemistry, one may come across many equations interrelating physical quantities. Therefore, any equation representing between physical quantities must be dimensionally correct. Thus quantities showing different dimensions cannot be equated. Let us consider a simple equation between two quantities, x and y then x = y. If the equation to be true, not merely representing a mathematical statement, then it can be said that x has a certain numerical value so also Y should have the numerical value. If the above equation is to be physically meaningful or dimensionally consistent, then the quantities must be true regardless of the set of units adopted for measurement. In other words, if the dimension of x is not equal to the dimension of γ,

If x has the dimension of length B has the dimension of time in the CGS system, A is expressed in meters and B is in seconds. Change to the SI system would result in a change in the numeral value of A because of the change of the length scale but there would be no change in the value of therefore, it is important to note that if the equal were to be numerically correct in one system of units. It could not be correct in the other system. Physical equations whose correctness do not change depend on a particular change of units are caused dimensionally consistent.

Above of course does not prove that it is correct. However, dimensionally consistency can prove that it is incorrect. It is derived entirely from dimensionally consistent equations. Checking each step of derivations of dimensionally consistency is a useful way to look for errors.

Handling of numbers

When a accumulation of large scientific data is in hand it is necessary to know the techniques for handling numbers associated with measurements.

Scientific notation

It is obvious from the fact that, in chemistry, we deal with numbers that are extremely large or extremely small. For example 1 mole or 1 g of Hydrogen element, there are approximately 602,300,000, 000,000,000,000,000, hydrogen atoms thus each hydrogen atom has mass of only 0.0000000000000000000000016.

These numbers are cumbersome to handle and it is easy to make mistakes when using them in arithmetic computation.

Let us consider the following multiplication

0.0000000056 x 0.00000000048 = 0.000000000000000002688.

In such cases it easy to add or miss one more after the decimal point. To handle these huge numbers, or very small numbers, we use a system called scientific notation. Regardless of their magnitude, all these numbers can be expressed with the formula $N \times 10^X$ where N is the number between 1 and 10, 'x' is an exponent that can be +ve or –ve integer. Any number expressed in this way is said to be written in scientific notation.

Suppose if we are given a certain number and asked it in scientific notation. This process amounts to finding 'n'. We count the number of places that the decimal point must be moved to give the number N. (It will be between 1 and 10). If the decimal point has to be moved to the left, then n is a positive integer. It has to be moved to the right, n is a negative integer.

Example: 1) Express 654.762 in scientific notation

$$654.762 = 6.54762 \times 10^2$$

It should be noted that the decimal point is moved to the left by two places and n=2

2) Express 0.00 000772 in scientific notation

$$0.00000772 = 7.72 \times 10^{-6}$$

Note that the decimal point is moved to the right by six places and n= -6

Significant figures

When we are measuring the quantity under investigation, it is impossible to measure exact amount of quantity. Therefore, for every value we measure, we must indicate the margin of error in a measurement by clearly indicating the number of significant figures, which are the meaningful digits in a measured or calculated quantity. When significant figures are concerned the last digit is found to be uncertain. The amount of this uncertainty depends on the particular measuring device.

Example:- (1) we represent the volume in 6ml as 6.0 ml.

In this case only one significant number that is '6' this is uncertain by -1 or +1. The actual value is some where between 5.9 and 6.1. We can further improve the measuring device and get accurate values.

(2) 845 has three significant numbers.

1.234 has four significant numbers.

Rules for using significant numbers

1. Any digit that is not zero is significant.

Example :- 645cm, has three significant figures.

1.834, has four significant figures.

2. The zero between non zero digits are significant.

Example :- 806 has three significant figures.

408031g, contains six significant numbers.

3. Zeros to the left of the first nonzero digits are not significant. The reason for these zeros is to indicate the decimal point

Example :- 0.08 mL contains one significant figure

0.000389 contains three significant figures.

4. If a number is greater than 1, then all the zeros written to the right of the decimal point count as significant figures.

Example :- 2.0 has two significant figures.

48.082g has five significant numbers.

0.4005g has four significant numbers.

5. The numbers that do not contain decimal points, the trailing zeros may or may not significant

Example :- 600 cm may have one, two or three significant numbers.

Practical exercise

1. Determine the number of significant figures in each of the measurements. a) 64mL b) 400 g c) 0.0430m^3 d) 6.4 x 10^4 molecules

Answer : a) Two b) four c)three d) two

Graphical representation of data

If a variable point p(x,y) moves under a given set of conditions, then the path traced by the point P(xy) is called the graph or locus of the point P. The equation that relates the variables x and y is called the equation of the graph of point. The graph can be represented by linear equation or simultaneous linear equation or a quadratic equation

1. Linear equation. In a linear equation one variable is an equation of the form ax + b = 0 or ax = C where a, b, c are constants (real numbers) and a ≠ 0 and x is variable

ii) An equation in the form of ax + by + c = 0 or ax + by = c where a, b, c are constants(real numbers) and a ≠ 0, b ≠ 0 and x and y are variables is called a linear equation with two variables.

Example :- Graphical representation of plant growth with time.

Plant height (in cm)	10	20	30	40	50	60	70	80	90	100
Time (hours)	10	20	30	40	50	60	70	80	90	100

The above data can be graphically represented in many ways.

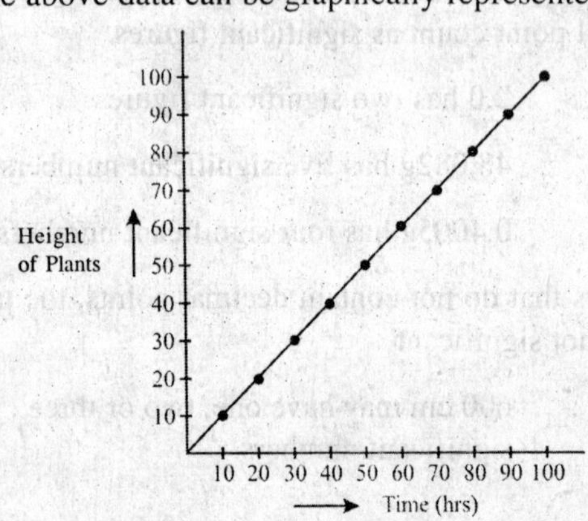

Line diagram

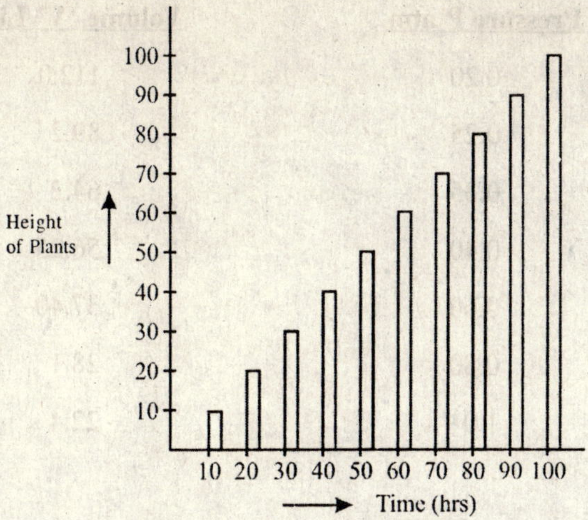

Bar diagram

2. Graphical representation of a typical elution profile by column chromatography of protein solution.

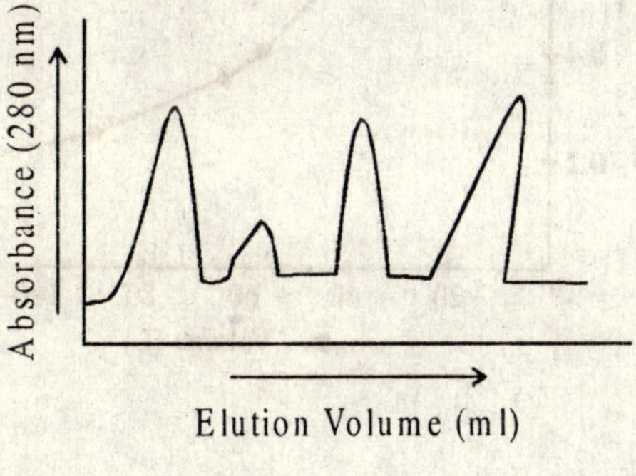

Conclusion : Each peak corresponding to one protein present in a given sample.

3. Effect of pressure on the volume of a gas at constant temperature.

Pressure P atm	Volume 'V' Liter
0.20	112.0
0.25	89.2
0.35	64.2
0.40	56.25
0.60	37.40
0.80	28.1
1.0	22.4

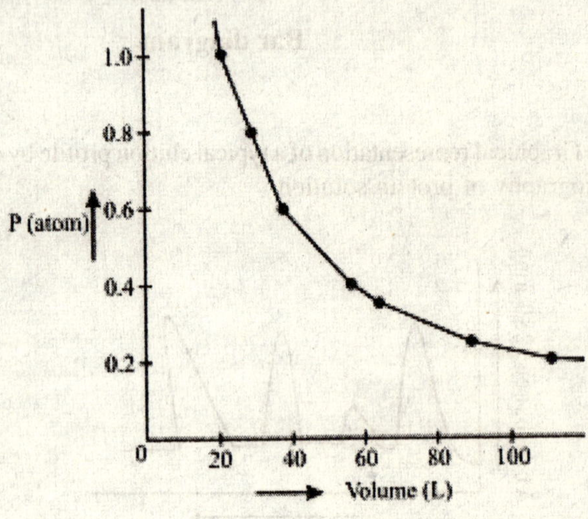

A series of experimental results measured is used to calculate the mean deviation. This shows agreement between the series of results that is measured. This is calculated by determining the arithmatical mean of the results are then evaluating the deviation of each individual measurement from the mean, are finally dividing the sum of the deviations regardless of sign by the number of measurements. The quantity of relative mean deviation is evaluated by dividing the mean deviation by the average of mean value. This can be expressed either in percentage or in parts per thousand. Generally, when a large body of the data is collected, it becomes necessary to express in terms of standard deviation which is normally accepted. The Standard deviation is nothing but a deviation of a population of observations. This is also called the r.m.s. deviation as it is the square root of the mean of the sum of squares of the difference between the values and the mean of those values. This is generally expressed mathematically as given below is of particular value in connection with the normal distribution.

Example: Let us consider the percentage of constitution X in a chemical compound XY percentages of X is found to be: 48.32, 48.36, 48.11, and 48.38%. We can calculate the mean deviation and the relative mean deviation as follows. The given data can be written as follows:

Results of the Experiment	**Deviation**
48.32	48.28 – 48.32 = 0.04
48.36	48.28 – 48.36 = 0.08
48.23	48.28 – 48.23 = 0.05
48.11	48.28 – 48.11 = 0.17
48.38	48.28 – 48.38 = 0.10
241.40	**Mean = 0.09**

$$\text{Mean} = \frac{241.40}{5} = 48.28$$

$$\text{Relative mean deviation in (\%)} = \frac{0.09 \times 100}{48.28}$$

$$= 0.19.$$

This example can lead to formation of an equation as follows.

Let us consider a series of experimental observations made on a system.

Let the number of observations be 'n' it will be arranged in the ascending order of magnitude:

$x_1, x_2, x_3,$ —— $x_{(n-1)}, x_n.$

Let $\bar{x}$ be the arithmatic mean (simply is called mean). This is given by

$$\bar{x} = \frac{x_1 + x_2 + x_3 + ---- + x_{(n-1)} + x_n}{n_1 + n_2 + n_3 + ---- + n}$$

$$\bar{x} = \frac{\sum_{i=1}^{n} x_i}{\sum_{i=1}^{n} n_i}$$

Let 's' be the standard deviation. This gives the spread of the values that are measured efficiently. The standard deviation is defined mathematically as

$$\sigma = \sqrt{\frac{(x_1 - \bar{x})^2 + (x_2 - \bar{x})^2 + ------ + (x_n - \bar{x})^2}{(n-1)}}$$

It may be noted here that in the above equation (n – 1) is used instead of 'n' value the number values is small. The above equation can be written in a compact manner as

$$\sigma = \sqrt{\frac{\sum_{i=n}^{n}(x_i - \bar{x})^2}{(n-1)}}$$

Variance : This is another important parameter in any branch of science specially in analytical chemistry. This is simply the square of the standard deviation (s). The variance is denoted by (small either) s. This is defined by

$$s = \sigma^2 = \sqrt{\frac{\sum_{i=n}^{n}(x_i - \bar{x})^2}{n-1}}$$

Correlation Coefficient

This is called also as coefficient of variation (C.V.). This is the most important factor which gives the more accurate measure of the precision. Then CV is defined by

$$C.V. = \frac{s \times 100}{\bar{x}}$$

Example : Analysis of a sample of iron ore gave the following percentage value for the iron content:

7.078, 7.210, 7.120, 7.090, 7.160, 7.140, 7.070, 7.180, 7.110. Calculate the mean and standard deviation and coefficient of variation for the values.

Average (Mean) Deviation and Standard Deviation

It has been found that, when a quantity is measured with the highest degree of exactness with the instrument method and observes, there is always an error in the measurement of data. The results of the measurement show that the successive determinations differ among themselves to greater or lesser extent. In such cases, an average or mean value is accepted. This is the most probable value of that measurement made on an observation. The most probable value may be true value in some cases as the difference may be small. In some other cases, it may be large. It is, therefore of interest into the factors which affect and control the faithfulness of the chemical analysis.

In any determination or measurement, errors are bound to happen. The errors could be of types like (1) Absolute error or (2) Relative error. The absolute error of a determination or measurement is the difference between the observed or measured value and the true or most probable value of the quantity measured then, the absolute error is a measure of the accuracy of the measurement.

The relative error is calculated when an absolute error is divided by the true or probable value, either in percentage or in parts per thousand. It may be noted that the true or the absolute error can not be established experimentally. It is generally concluded that the result must be compared with the most probably value. It is known that the atomic weights of the elements in the periodic table have been determined with the greatest care. Hence, with pure substance the quantity will ultimately depend upon the atomic weights of the constituent elements. In this case the accuracy of the measured value is very high. If several analysis determine the some constituent in the same given sample by different methods, the most probable value, which is usually the average, which can be deduced from their results. In both cases the establishment of the most probable value involves the application of statistical methods and the concept of precision.

2

THE ATOMIC STRUCTURE

Introduction

What constitutes the matter?

Man with his inquisitive mind has been studying over the ages to understand the constitution of matter. In the 5th century B.C. Democritus (the Greek Philosopher) expressed the belief that all matter (Solids, liquids and gases) consists of an all tiny, individual particles, what he called atoms (meaning indivisible). However, his idea was not accepted by his contemporaries like Plato and Aristotle. Early scientific investigations of the matter provided the support of the notion of "atomism" and slowly give rise to the modern definition of elements and compounds. It was John Dalton (an English scientist) in 1808 who formulates a precise definition of the indivisible building blocks of matter what we call now atoms. His work marked modern era of chemistry. The hypothesis about nature of matter is known as Dalton's atomic theory.

The Structure of the Atom

What is an atom?

We can define an atom based on the hypothesis of Dalton's atomic theory. According to his hypothesis, an atom is the basic unit of an element that can enter into chemical combination. However, a series of scientific investigations over the years led to demonstrate that atoms really possess internal structure. This means that the atoms are made up of some smaller particles these are now termed as sub atomic particles. These subatomic particles are electrons, protons and neutrons.

The Electron

Many scientists in the 1950's were involved in the study of radiation which is associated with the emission or transmission of energy through space, in the form of waves, the scientific data gained in this has contributed gradually to an understanding of atomic structure.

J.J. Thomson (an English physicist) used a cathode ray tube and his wave theory of electromagnetic theory to determine the ratio of electric charge to the mass of an individual electron. Thereafter in a series of experiments carried out between 1908 and 1917, R.A. Micok succeeded in measuring charge of each electron. He found the charge of an electron to be 1.600 x 10^{-19} C.

From the data, the calculated the mass of an electron.

Mass of an electron = Charge / Mass

$$= \frac{-1.16022 \times 10^{-19C}}{-1.76 \times 10^{3}\, C/g}$$

$$= 9.10 \times 10^{-19}\, g$$

The mass of an electron is an exceeding small mass.

2. Discovery of proton :

In the early 1900's it became that the atoms contain an equal number of +ve charged particles an opposite to the equal number of electrons which are –vely charged particles. J.J. Thomson, therefore proposed that an atom could be thought of +ve sphere of matter in which electrons are embedded like raises in a coke.

Rutherford who carried out a series of experiments using the foils of gold and other metals as targets for $\propto$ – particles, scattering. He was later able to explain the result of his scattering experiment in terms of a new model for the atom. According to him, most of the space in the atom must be empty because majority of the particles passed through the gold foil with little or no deflection. This led to the conclusion that +ve charges that are all concentrated in the nucleus.

What is nucleus? This is a dense central part of the atom. The +vely charged particles in the nucleus are called protons. It was in separate experiment from that each proton carries the same quantity of charge of an electron, and has a mass of 1.67262×10^{-24}g which is about 1840 times the mass of the oppositely charged electron. What is the scientists perception of an atom?

The mass of a nucleus of an atom constitutes most part of the mass of the entire atom, in an incredibly small volume of about 1×10^{-13} of the volume of the entire atom. The concept of atom is very useful, but it could not be inferred that atom have well defined boundaries or surfaces. However, the outer regions of atom are really relatively "fuzzy".

The Neutrons:

The Rutherford model of atomic structure left one major problem unsolved. However Rutherford and other suggested that there must be another type of subatomic particle in the atomic nucleus. The proof for this particularly was provided by James Chadwick (an English Physicist) in the year 1932, when Chadwick bombarded beryllium foil with the particles a very high energy radiation was emitted that was similar to γ – rays. An analysis of the experimental results showed that the γ - rays actually consisted of a third type of subatomic particles, which Chadwick called neutrons which then proved to be electrically neutral particles having a mass slightly greater than that of protons.

There are other subatomic particles but the electrons, the protons and the neutrons are the only three fundamental components of the atom that are involved in chemistry.

The following table shows the masses and the charges of these three elementary p[articles.

Mass and charge of elemental particles

Particle	*Mass (z)*	*Charge*	*Unit*
Electron	9.10939×10^{-24}	-1.6022×10^{-19}	- 1
Proton	1.67262×10^{-24}	$+1.6022 \times 10^{-19}$	+1
Neutron	1.67493×10^{-24}	0	0

1.2 Rutherford's Gold Foil Experiment: Discovery of the Nucleus

In the year 1911, the *nucleus* was accidentally discovered by Ernest Rutherford. When he along with his co-workers Marsden and Geiger - was studying the scattering of α - particles (He^{2+} ions) by a thin gold foil, were observed the following features: (Fig. 2.1):

- Majority of the α - particles went through the foil undeflected.
- A good number of the particles got deflected through varied angles.
- Very few particles got *reflected back* by the foil.

The third observation shocked Rutherford. He said, "*It was unbelievable. It was as incredible as if you fire a* 15 *inch shell at a piece of tissue paper and it came back and hit you!*". However, once the shock was over, he analyzed the results and suggested the following:

- The atom is largely hollow.
- Each atom has a *centrally located positively charged body called the nucleus.*
- The nucleus is very small in size compared to the total volume of the atom but almost the entire mass of the atom is concentrated in it.

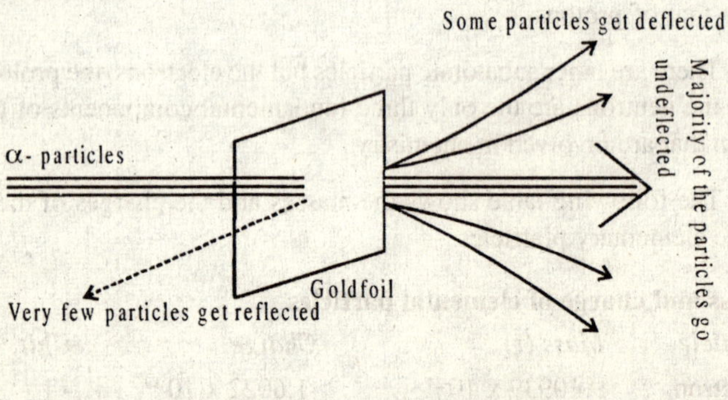

Fig. 2.1 The gold foil experiment

1.2.1 Rutherford's Planetary Model of Atom:

Combining the above ideas with the existence of electrons, Rutherford proposed a model for the atom. According to him, the atom has a centrally located positively charged nucleus. The electrons revolve around the nucleus in circular orbits in the same way as the planets revolve around the sum (Fig 2.2 a)

1.2.2 Limitations of the Planetary Model of Atom:

Rutherford's planetary model of atom suffered the following limitations.

- Any charged body revolving around another charged body should emit energy in the form of electromagnetic radiation. Hence, the revolving electron should lose energy steadily, take a spiral path and collapse into the nucleus (Fig. 2.2 b)
- This model could not explain the emission spectra of elements (Ref. p.4)

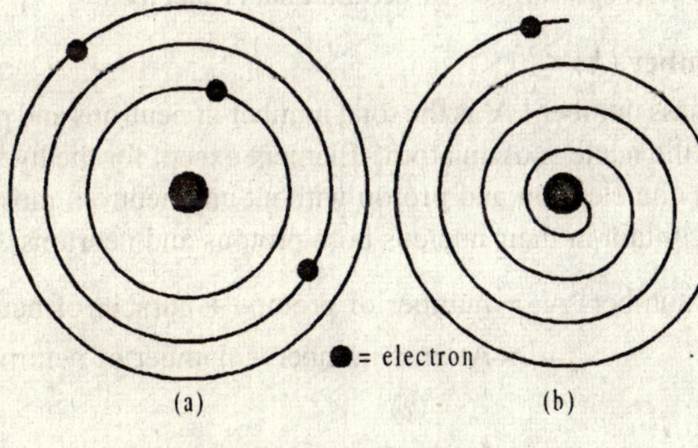

Fig. 2.2 a) Planetary model of atom
b) Instability of the planetary model

Composition of Nucleus

According to modern views, nucleus is composed of protons and neutrons which are together known as nucleons since, each proton has a single positive charge, the number of protons in the nucleus of a given element is equal to the nuclear charge which inturn is equal to atomic number.

The attractive and repulsive forces exist in the nucleus. The attractive forces, which tend to stabilize the nucleus exist between neutrons and protons and are due to energy exchange between these particle. The repulsive forces which decreases the stability of the nucleus, exist between protons since like charges repel each other. The nucleus must reach an equilibrium between these forces in order to become stable.

Atomic number(z)

Atomic number is the number of protons in the nucleus of each atom of an element. In the neutral atom, the number of protons is equal to the number of electrons so, the atomic number also indicated the number of electrons present in the atom.

Example: 1. The atomic number of nitrogen is 7. This means that each neutral nitrogen atom has 7 protons and 7 electrons.

2.The atomic number of sodium is 11. This means that each neutral atom of sodium has 11 proton and 11 electron.

Mass Number (A)

The mass number ' A' is the total number of neutrons and protons present in the nucleus of an atom. Element except for the hydrogen, which has one electron and proton without any neutron, most other elements contain in their nucleus both protons and neurtons.

Mass number 'A' = number of protons + number of neutrons

= Atomic number (z) + number of neutrons.

∴ number of neutrons = mass number 'A' – Atomic number (z)

Atoms that have the same atomic number but different mass number are called isotopes.

The accepted version of denoting atomic number and mass number of an atom as follows.

$^{A}_{Z}X$ Where A = Mass number, Z = Atomic number.

Electromagnetic spectrum

All forms of light, visible, infrared, x – rays, gamma – rays are regarded as electromagnetic radiations. Each kind of radiation extends over a range of wavelength. For example, the visible light has a wavelength of 350nm (violet) to 750nm (red). The electromagnetic radiations spectrum ranges from gamma rays to radio waves as shown in fig. 2.3.

0.01 to zero	001 to 150 A^0	0 – 1 t0 150 A^0	150 to 3800 nm	3800 A^0 to 7600 A^0	7600 to 6 x 10^6 A^0	3 x 10^6 to 3 x 10^9 A^0	3 x 10^7 to 3 x 10^{14} A^0
Cosmic rays	r – rays	X – rays	Ultra violet rays	Visible rays	Infra red waves	Micro waves	Radio waves

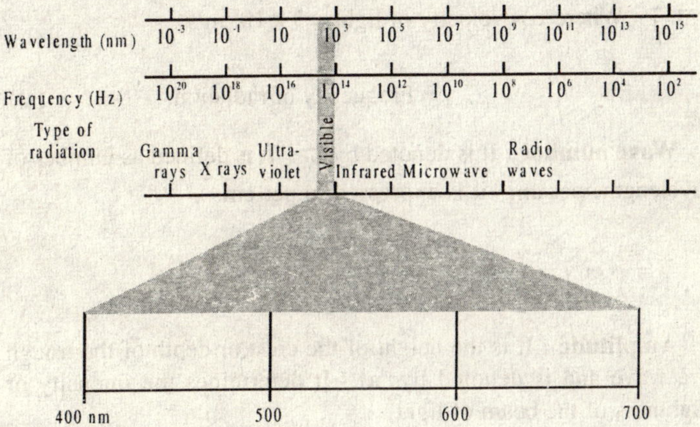

Fig. 2.3 Electromagnetic Spectrum.

A wave is sort of disturbance which originates from, some vibrating source and travels outwards as a continuous sequence of alternating crests and trough. Every wave has five important characteristics, namely, Wavelength (λ), Frequency (γ), Velocity (c), Wave number ($\bar{\nu}$) and amplitude (a).

X – rays, γ - rays, visible light, infrared light and radio waves are called electromagnetic radiations because similar waves can be produced by a moving charged body in a electric field and a magnetic field. These radiations do not require the medium for their propagation. The properties of the electro magnetic radiation are as follows:

Wavelength (λ) : The distance between two neighbouring troughs or crests is known as wavelength. It is denoted by λ. It is expressed in Angstrom units (A^0) or nanometer (nm)

$$1A^0 = 10^{-8} \text{ cm and } 1\text{nm} = 10^{-9}\text{m.}$$

Frequency (γ) : The frequency of a wave is the number of times a wave passes through a given point in a medium in one second. It is denoted by (γ) and is expressed in cycles per second.

Velocity : Velocity is the distance traveled by a wave in one second. It is denoted by C and is expressed in cm sec^{-1}. $c = \gamma\,\lambda$ or

$\lambda = \dfrac{c}{\gamma}$. Where c = velocity of light = 3×10^8 m/s

γ = Frequency of radiation.

Wave number : It is denoted by $\bar{\nu}$. It is defined as number of wavelengths per cm. it is expressed in per cm.

$$\bar{\nu} = \frac{1}{\lambda} \text{ or } \bar{\gamma} = \frac{\gamma}{C}$$

Amplitude : It is the height of the crest or depth of the trough of a wave and is denoted by 'a'. It determines the intensity or brightness of the beam of light.

Atomic spectrum

Atoms of elements can be made to emit energy by subjecting them to electric discharge or by heating them strongly. The emitted radiations can be resolved by a prism into a spectrum. The emission spectrum thus obtained is discontinuous. It consists of lines of different colours. The rest of the spectrum is dark, hence it is called line spectrum.

If the atom gains energy, the electron passes from a lower energy level to a higher energy level, the energy is absorbed, that means energy of a specific wavelength is absorbed. Consequently, a dark line will appear in the spectrum. This dark line constitutes the absorption spectrum.

If the atom loses energy, the electron passes from higher to a lower energy level, energy is released and spectral line of specific wavelength is emitted. This line constitutes the emission spectrum.

Types of emission spectra

Continuous spectra : When white light from any source such as sun is analyzed by passing through a prism, it splits up into seven different colours from violet to red. These colours are so continuous that each of them merges into the next. Hence, the spectrum is called continuous spectrum.

Line spectra : When the light emitted by the gas discharge tube is resolved by spectroscope, bright lines are observed which are separated by dark spaces. Such spectrum is line spectra.

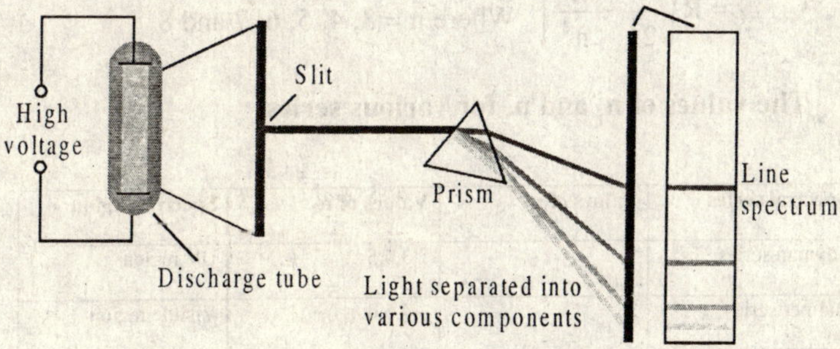

Fig.: **2.4** Emission spectrum of hydrogen

When an electric discharge is passed through hydrogen gas at low pressure (0.001 atm), the gas emits a rose coloured light. When a beam of this light is dispersed by a prism, it is found to consist of series of lines in the UV, visible, and IR regions. This series of lines is known as line spectrum of hydrogen or Atomic spectrum of hydrogen. Each line corresponds to a particular frequency. These lines could be grouped into a definite number of spectral series. There are five series, which are known as (i) Lymann series ii) Balmer series (iii) paschen series (iv) Brackett series (v) Pfund series.

The wavelength of all these series can be expressed by a single formula

$$\frac{1}{\lambda} = \bar{\gamma} = R\left[\frac{1}{n_1^2} - \frac{1}{n_2^2}\right]$$

where $\bar{\gamma}$ = Wave number,

λ = wavelength

R = Rydberg constant (109678 cm^{-1})

The Lyman series lies in the UV region, the Balmer series lies in the visible region and the remaining three series in the infrared region.

The Balmer series contains four prominent light designated as H, H_β, H_γ and H_δ. The wave number of each spectral line of Balmer series is given by

$$\bar{\gamma} = R\left(\frac{1}{2^2} - \frac{1}{n^2}\right) \quad \text{Where n = 3, 4, 5, 6, 7 and 8}$$

The values of n_1 and n_2 for various series.

Spectral series	Values of n_1	Values of n_2	Spectral region
Layman series	1	2,3,4,5...	UV region
Balmer series	2	3,4.5.6...	Visible region
Paschen series	3	4,5,6,7...	Infrared region
Bracket series	4	5,6,7,8	Infrared region
Pfund series	5	6,7,8	Infrared region

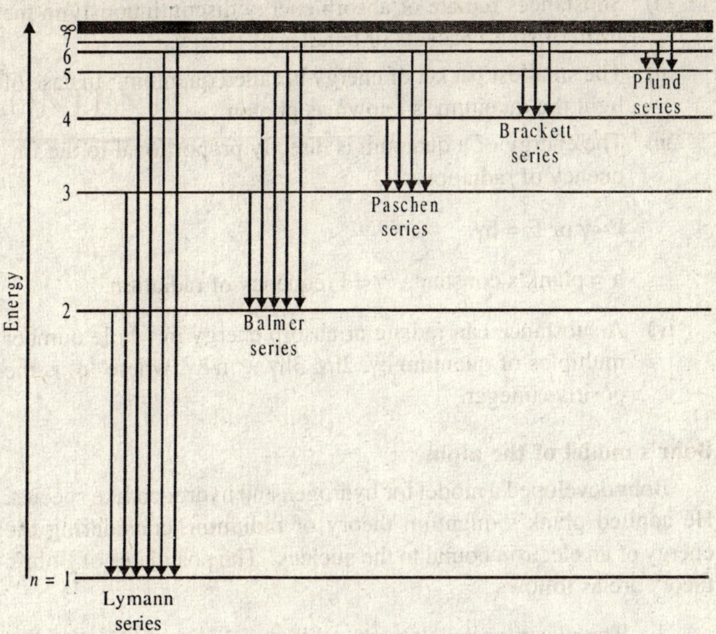

Fig. 2.5. Electronic transitions corresponding to the spectral series of hydrogen.

Plank's theory of radiation

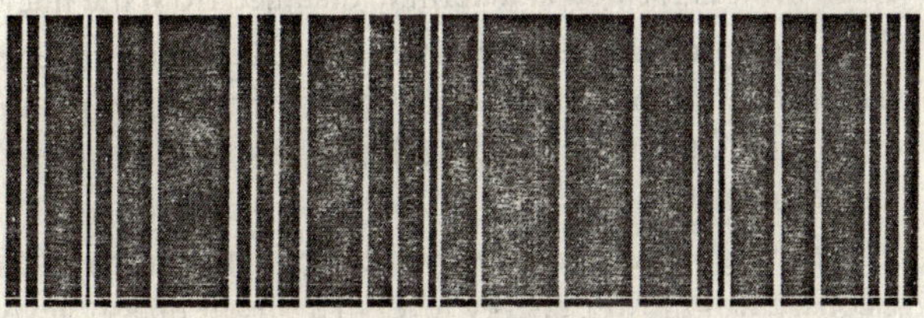

Line spectrum of hydrogen atom

PLANKS QUANTUM THEORY

The quantum theory of radiation was proposed by max-plank in 1904. According to his theory, absorption or emission occurs in discrete units or bundles called quanta and not in continuous manner. The main points of quantum theory are.

i) Substances radiate or absorb energy discontinuously in the form of small packets or bundles of energy.

ii) The smallest packet of energy is called quantum. In case of light the quantum is known as photon.

iii) The energy of a quantum is directly proportional to the frequency of radiation.

$E \propto \gamma$ or $E = h\gamma$

h = plank's constant. γ = Frequency of radiation.

iv) A substance can radiate or absorb energy in whole number multiples of quantum $h\gamma$, $2h\gamma$, $3h\gamma$.... n $h\gamma$, where 'n' is the positive integer.

Bohr's model of the atom

Bohr developed a model for hydrogen and hydrogen like species. He applied plank's quantum theory of radiation, considering the energy of an electron bound to the nucleus. The postulates of Bohr's theory are as follows.

1. The electrons revolve around the nucleus in a fixed closed orbits known as stationary states, but no radiation is emitted. The force of attraction between the nucleus and an electron is equal to the centrifugal force of the moving electron.
2. Each orbit is located at a definite distance from the nucleus.
3. The energy of an electron depends on the orbit it occupies. The farther an orbit from the nucleus, the greater is the energy associated with it. The orbit nearest to the nucleus has the least energy.
4. If the energy is supplied to an electron, it may jump instantaneously from lower energy level to a higher energy level by absorbing one or more quanta of light energy. This new state of an electron is called excited state. Since the excited state is less stable, atom loose energy and come back to the ground state.

   Energy released or absorbed in an electron jump, (ΔE) is given by $\Delta E = E_2 - E_1 = h\gamma$. Where E_1 and E_2 are energy levels, h is plank's constant, γ is frequency of radiation.

5. Of the finite number of circular orbits around the nucleus an electron can revolve only in those orbits whose angular momentum (mvr) is an integral multiples of factor, $mvr = \frac{nh}{2\pi}$

   Where m = mass of the electron, v = velocity of electron, n = orbit number in which electron is present, h = planks constant, r = radius of the orbit.

Merits of Bohr's theory

1. The frequencies of the spectral lines determined experimentally are in excellent agreement with the values calculated from Bohr's formula.
2. The experimental value of Rydberg constant 'R' for hydrogen, agrees with the value calculated from Bohr's theory.
3. The emission and absorption of hydrogen spectrum are explained nicely by Bohr's theory.
4. The radii and energies of the orbits in hydrogen atom as calculated on the basis of Bohr's theory are in good agreement with experimental values.

Demerits

1. Bohr's theory does not explain the spectra of atoms having more than one electron.
2. Bohr's theory failed to account for the effect of magnetic field (zee man effect) or electric field (stark effect) on the spectra of atoms or ions. When the source of a spectrum is placed in a strong electric or magnetic field, each spectral line further splits into a number of lines. This phenomenon was not explained by Bohr model.
3. Bohr model treated electron as particle only. However, De Broglie suggested that the electrons posses both particle as well as wave properties.
4. The postulates of Bohr, that electrons revolve in well defined orbits around the nucleus with well defined velocities is not in accordance with Heisenberg's uncertainty principle.
5. Atom is assumed to be planer.

6. The reason for quantization of energy is not presented.

Dual nature

Particle and wave nature of light can be explained by De Broglie equation. Light is considered as a particle because it exhibits the properties such as black body radiation and photoelectric effect. The wave nature is explained by the phenomenon like diffraction and interference can be explained on the basis of its wave character. Thus light is said to contain wave as well as particle nature. These studies were done by Einstein in 1905.

Louis De Broglie, in 1924 extended the idea of photons to material particles such as electron and he proposed that matter is also has dual characteristics as wave and a particle.

Derivation of De Broglie equation

The wavelength of a wave associated with any material particle was calculated by analogy with photon. As particle of matter in motion describes a wave. Thus a moving electron is associated with a definite wavelength, given by $\lambda = \frac{h}{mu} = \frac{h}{p}$ where λ wavelength m = mass of electron, u = velocity of moving electron, p = momentum of the electron.

The above equation is called de – Broglie's equation. The wave nature of the electron has been confirmed by the diffraction of electron beam by crystal lattice. In view of this fact Bohrs theory was failed.

According to Einstein's mass energy relationship which is given by $E = mc^2$.

Equating this energy with the energy of the photon, we get $h\gamma = mc^2$.

We know that $\gamma = \frac{c}{\lambda}$

$$\therefore \frac{hc}{\lambda} = mc^2$$

$$\text{Hence } \lambda = \frac{h}{mc}$$

By substituting c by the velocity of the electron u we have

$\lambda = \dfrac{h}{mu}$ This is Debroglie's equation.

Worked Examples :

1. Calculate the wavelength of the particle moving with a speed 130 miles per hour or 58 m/s, the mass of the ball is 6.0×10^{-2} kg.

$\lambda = \dfrac{h}{mu}$ $\quad h = 6.63 \times 10^{-34}$ Js

$m = 6.0 \times 10^{-2}$ kg

$u = 58$ m/s.

$$\lambda = \frac{6.63 \times 10^{-34}}{6.0 \times 10^{-2}\ \text{kg}} = 1.9 \times 10^{-34}\ \text{m.}$$

2. Calculate wavelength (nm) of a H – atom moving with the speed of 7.0×10^2 m/s.

$\lambda = \dfrac{h}{mu}$ $\quad h = 6.63 \times 10^{-34}$

$m = 1.674 \times 10^{-27}$

$u = 7.0 \times 10^2$

$$\lambda = \frac{6.63 \times 10^{-34}\ \text{JS}}{1.674 \times 10^{-27} \times 7.0 \times 10^{2}\ \text{m/s}} = 56.6\ \text{nm.}$$

3. Calculate the frequency of a radiation having wavelength 600nm.

$c = \gamma\lambda$ $\quad c = 3 \times 10^8$ m/s $\quad \lambda = 600\text{nm or } 600 \times 10^{-9}$ m

$$\gamma = \frac{c}{\lambda}$$

$$\gamma = \frac{3 \times 10^8 \text{ m/s}}{6 \times 10^{-7} \text{ m}} = 5 \times 10^4 {}_{-1}$$

4. A visible radiation has frequency 6×10^{14} cycle / Sec. Find out the wavelength of the radiation in nanometers.

$$c = \gamma\lambda$$

$$\lambda = \frac{C}{\gamma} \qquad c = 3 \times 10^8 \text{ m/s}$$

$$\gamma = 6 \times 10^{14} \text{ cycle/ sec}$$

$$\lambda = \frac{3 \times 10^8 \text{ m/s}}{6 \times 10^{14} \text{ cycles/sec}} = 0.5 \times 10^{-6} \text{ m}$$

5. Calculate the energy of a photon of light having frequency of 1.0×10^{15} s^{-1} ($h = 6.63 \times 10^{-34}$ J s^{-1}

$$E = h\gamma \qquad \gamma = 1.0 \times 10^{15} \text{ s}^{-1}$$

$$E = 6.63 \times 10^{-34} \times 1.0 \times 10^{15}$$

$$E = 6.63\ 10^{-19} \text{ J}$$

6. Calculate the wavelength of H_{∞} and H_{β} lines of hydrogen spectrum in the Balmer series. The value of Rydberg constant is 10.97×10^6 cm^{-1}

For H_{∞}, n = 3.

$$\overline{\gamma} = \frac{1}{\lambda} = R\left(\frac{1}{2^2} - \frac{1}{n^2}\right)$$

$$\frac{1}{\lambda} = 10.97 \times 10^6 \left(\frac{1}{2^2} - \frac{1}{3^2}\right)$$

$$= 10.97 \times 10^6 \times \frac{5}{36}$$

$$\lambda = \frac{36}{10.97 \times 5 \times 10^6}$$

$$\lambda = 656.3 \times 10^{-9} \text{ m or } 656.3 \text{ nm}$$

For H_β, n = 4.

$$\frac{1}{\lambda} = 10.97 \text{ x } 10^6 \text{ x} \left(\frac{1}{2^2} - \frac{1}{4^2}\right)$$

$$\frac{1}{\lambda} = 10.97 \text{ x } 10^6 \left(\frac{3}{16}\right)$$

$$\frac{1}{\lambda} = 10.97 \text{ x } 10^6 \left(\frac{1}{2^2} - \frac{1}{3^2}\right)$$

$$\lambda = \frac{16}{10.97 \text{ x } 10^6 \text{ x } 3} = 486.1 \text{ x } 10^{-9} \text{ m} \quad \text{or } 486.1 \text{ nm}$$

7. Calculate the wave number and wave length of the first spectral line of the Lyman series of hydrogen spectrum. (R = 10.97 x 10^6 m^{-1}).

For first line $n_1 = 1$ & $n_2 = 2$

$$\bar{\gamma} = R\left[\frac{1}{n_1{}^2} - \frac{1}{n_2{}^2}\right]$$

$$\bar{\gamma} = 10.97 \text{ x } 10^6 \left[\frac{1}{1^2} - \frac{1}{2^2}\right]$$

$$\bar{\gamma} = 10.97 \text{ x } 10^6 \text{ x } \frac{3}{4}$$

$$\lambda = 8.2275 \text{ x } 10^6 \text{ m}^{-1}$$

Since λ and $\bar{\gamma}$ are related as $\frac{1}{\lambda} = \bar{\gamma}$

$$\frac{1}{\lambda} = 8.2275 \text{ x } 10^6 \text{ m}^{-1}$$

$$\lambda = \frac{1}{8.2275 \text{ x } 10^6} = 121.5 \text{ x } 10^{-9} \text{ m} \quad \text{or} = 121.5\text{nm}$$

8. Calculate the wavelength associated with (a) An electron moving with a velocity of $1 \times 10^6 m\ s^{-1}$ (b) a cricket ball of mass 160g moving with a velocity of $42\ m\ s^{-1}$.

a) Mass of the electron $= 9.109 \times 10^{-31}$ kg.

Momentum of the electron = m x v

$$(p) = 9.109 \times 10^{-31} \times 10^6$$
$$= 9.109 \times 10^{-25}\ Kg\ ms^{-1}$$

$$\lambda = \frac{h}{p} = \frac{6.626 \times 10^{-34}}{9.109 \times 10^{-25}}$$

$$= 7.274 \times 10^{-10} m$$

b) Momentum of the ball. $p = 0.16 \times 42\ kg\ ms^{-1}$

$$= 6.72\ kg\ ms^{-1}$$

$$\lambda = \frac{h}{p}$$

$$= \frac{6.626 \times 10^{-34}}{6.72}$$

$$= 9.86 \times 10^{-35} m$$

9. Calculate the momentum of a photon of radiation of wavelength $500A^0$. ($h = 6.626 \times 10^{-34}$ JS)

$1\ A^0 = 1 \times 10^{-10}$ m

According to De Broglie, $\lambda = \frac{h}{p}$

$$P = \frac{h}{\lambda}$$

$$P = \frac{6.626 \times 10^{-34}}{500 \times 10^{-10}}$$

$$P = 1.3252 \times 10^{-26}\ kg\ m\ s^{-1}$$

10. What is the wave length of a photon emitted during a transition from $n_1 = 5$ to $n_2 = 2$ state in H atom.

$$\Delta E = R_H\left[\frac{1}{n_1^2} - \frac{1}{n_2^2}\right] \qquad RH = 2.18 \text{ X } 10^{-18}$$

$$= 2.18 \times 10^{-18}\left[\frac{1}{5^2} - \frac{1}{2^2}\right]$$

$$\Delta E = -4.58 \times 10^{-19} \text{ J}$$

$$\text{Since } E = h\gamma \quad \text{or } \gamma = \frac{\Delta E}{h}$$

$$\lambda = \frac{6.63 \times 10^{-34} \times 3 \times 10^{8}}{-4.58 \times 10^{19} \text{ J}}$$

$$\lambda = 4.34 \times 10^{-7} \text{ m}$$

Origin of the spectral lines of hydrogen atom

Bohr explained the emission spectra of hydrogen. When an electron occupies the orbit n = 1, that is the most stable state if hydrogen atom. The atom in this state has minimum potential energy and is said to be in the ground state. This refers to the lowest energy state of a system. The stability of the electron diminishes form n = 2, 3, ... and each of this is called as exited state, which is higher in energy than the ground state.

When energy is supplied to hydrogen gas by passing an electric discharge, the molecules absorb energy and split up into atoms. The atoms also absorb energy. Different atoms absorb different quanta of energy. As a result, the electrons present in the ground state of atoms jump into higher orbits. The position taken up by the electrons in different atoms present in the gas are different. When the electron jump back to lower orbits, the quanta of energy absorbed earlier is given out. Thus energy appears in the form of radiations of definite frequencies or wavelength. This is observed as lines in emission spectrum. For example, lines in the Lyman series are formed when an electron jumps from higher orbits to orbit number one (n = 1). Similarly the Balmer light is produced when the electron jumps from outer orbits to orbit number two. Similarly, the lines in Paschen series, Bracket series and fund series are formed by the transitions of electron from outer orbital to the orbital number 3, 4, and 5 respectively.

Distinction between particle and wave

It is found that the matter waves are distinctly different from electromagnetic radiation waves. The speed of these waves are not same as that of light. It is very small compared with light waves. Matter waves cannot be radiated in empty space. The wavelength of the matter waves are generally very small as compared to the wave length of electromagnetic waves.

Heisenberg's uncertainty principle

It states that "it is impossible to measure simultaneously, the position and momentum of a small microscopic moving particle with absolute accuracy or certainty". This means, if an attempt is made to measure any one of these two quantities with higher accuracy, the other becomes less accurate.

The product of the uncertainty in position (Δx) and the uncertainty in the momentum ($\Delta P = m \Delta V$) where m is the mass of the particle and ΔV is the uncertainty in velocity) is equal to or greater than $\frac{h}{4\pi}$ where h is the plank's constant.

Thus, the mathematical expression for the Heisenberg's uncertainty principle is simply written as $\Delta x . \Delta p \geq \frac{h}{4\pi}$.

SCHRODINGER'S EQUATION

In 1926, an year before Heisenberg proposed the uncertainty principle, Erwin Schrodinger came out with an equation which can be considered the fundamental equation of quantum mechanics. Assuming a minute particle to be a standing wave, he derived a mathematical equations which are second order differential equations can be used in solving problems related to the motion of minute particles such as the electrons. Schrodinger's equation may be presented in a simple form as $H \psi = E \psi$

Here, H (read as H capped) is a mathematical operator called the Hamiltonian operator – usually a combination of a number of double differential operators. ψ is called the *state function* of the system under study or the *wave function,* as it is a measure of the amplitude of the wave associated with the minute particle.

What are Functions and Operators?

In science, many a times, we study the change in a certain variable (property) of a system with respect to other variables. For example, we may study the change in the pressure, P of a gas with its volume, V and the absolute temperature, T. Such studies yield relationships of the type.

$Y = f(x_1, x_2, x_3 \ldots\ldots.)$

Where $x_1, x_2, x_3\ldots\ldots$ Are independent variables on which y depends. We say that y is a function of $x_1, x_2\ x_3\ldots\ldots.$

Functions can be classified into continuous and discontinuous functions. A continuous function has a value at all values of its independent variables. A discontinuous function does not have values for certain values of its independent variables.

A Mathematical operator is a symbol written usually to the left of a function to represent the mathematical operation that should be carried out on the function. For example, in sine x, x is the function and the symbol sine is the operator. Sin tells us to take the sine value of x. It should be borne in mind that sin x is not understood as sine multiplied by x. Similarly, in $\frac{d}{dx}[f(x)]$ $\frac{d}{dx}$ is an operator. It calls for differentiating the function f(x) with respect to (a regular procedure in calculus). A double differential operator is of the form $\frac{d^2}{dx^2}$.

It calls for differentiating the ensuing function twice. That is $\frac{d^2}{dx^2}[f(x)]$ is equivalent to $\frac{d}{dx}\left\{\frac{d}{dx}[f(x)]\right\}$.

1.8.1 Schrodinger's Equation for the Hydrogen Atom:

Hydrogen atom is made of a positively charged proton. The atom need not be flat as assumed in Bohr's model. If we take the nucleus as the origin of a Cartesian coordinate system, the electron can be present in the three dimensional space around the nucleus and its position can be given by the x, y, z coordinate values. Schrodinger's equation for a particle under a potential field is

$$\left(\frac{\partial^2}{\partial x^2}+\frac{\partial^2}{\partial y^2}+\frac{\partial^2}{\partial z^2}\right)\psi+\frac{8\pi^2 m}{h^2}(E-V)\psi=0 \qquad [1]$$

Where $\frac{\partial^2}{\partial x^2}$, $\frac{\partial^2}{\partial y^2}$ and $\frac{\partial^2}{\partial z^2}$ are partial double differential operators operating on the wave function, ψ. E and v are the total energy and the potential energy of the particle respectively. m is the mass of the particle and h is Planck's constant.

In the hydrogen atom, electron experiences a potential field due to the nucleus. The potential energy of electron due to the nucleus. The potential energy of electron due to this field is $-e^2/r$ where e is the charge on the electron and r is the distance between the nucleus and the electron. Substituting this potential energy expression in equation (1.33), we get the Schrodinger Equation for the hydrogen atom.

The Wave Function, ψ:

The wave function, is a measure of the amplitude of the standing wave associated with the particle under study. It does not signify the position of the particle. However, its squared value is of physical significance. According to Max Born, ψ^2 in a small volume element dτ, given by ψ^2 dτ is the probability of finding the particle in the volume element dτ. Note that this physical meaning of the wave function is in accordance with the uncertainty principle .

As the energy of the particle is fixed, there is a lot of uncertainty in the position of the particle and we can talk about its probable position only. Also, the idea that ψ^2 d τ predicts the probability of finding the particle in the volume element puts certain restrictions on the wave function.

Restrictions on ψ

As ψ^2 is the probability of finding the election, ψ has the following restrictions.

ψ and its first derivative $\left[\frac{d\psi}{dx}\right]$ should be continuous functions within the boundaries of the system.

The total probability of finding the particle in the entire system should be unity. That is,

$$\int_1^2 \psi^2 \, d\tau = 1$$

ψ should vanish, that is, become zero at the boundaries of the system.

Schrodinger's Equation is an Eigen Value Equation:

Differential equations like (1) can have infinite solutions. However, the restrictions on the wave function restrict the number of solutions. Equations for which infinite solutions are possible but only certain solutions are acceptable due to some restrictions are called eigen value equations. The acceptable solutions are called eigen values.

For example, when equation (1) is solved, a number of values for E emerge as solutions. These energies are the eigen values, corresponding to each eigen value, there exists a wave function called the eigen function. If the energies are labeled E_1, E_2 ...., the corresponding eigen functions can be labeled ψ_1, ψ_2, ...

Each eigen function differs from the others only in the values of one or more integral numbers called quantum numbers. In the case of hydrogen atom, eigen functions are characterized by three quantum numbers. [It is a general rule that the wave functions of any system will be characterized by as many quantum numbers as there are coordinates (dimensions)].

Orbitals

$\psi^2\, d\tau$ gives the probability of finding the electron in the small volume element $d\tau$. Now, for a particular energy value of the electron in the hydrogen atom, say E_n, the corresponding wave function is ψ_n. The square of this function in different volume elements around the nucleus can be computed. Since, the potential energy term V vanishes only at infinity ψ vanishes only at infinity. Hence, the probability of finding the electron of the hydrogen atom has a finite value even at large distances from the nucleus.

It is important to understand the difference between the terms 'orbit' and 'orbital'. The term 'orbit' which we encountered in Bohr's model of atom refers to a well defined circular path around the nucleus in which the electron moves. The term 'orbital' does not refer to any specified path of the electron. It refers to an imaginary volume around the nucleus where the probability of finding the electron is maximum.

While the concept of orbital is in accordance with the uncertainty principle the concept of orbit is not.

QUANTUM NUMBERS AND SIMPLIFICATION OF THE ATOM

Formal treatment of the structure of the atom using Schrodinger's equation can be presented using a simplified approach here. This approach uses the results of Schrodinger's equation - especially the quantum numbers that characterize the wave functions.

Combining the ideas of relativity with Schrodinger's equation, Dirac showed that the wave function corresponding to each energy state of an electron should involve another quantum number called the spin quantum number which can be associated with the spinning of the electron about its own axis. Thus there are four quantum numbers that characterize each energy state of the electron in the hydrogen atom.

The energy and the spatial distribution of an electron in a particular energy state are controlled by these four quantum numbers. An atom contains large number of shells and sub shells. These are distinguished from one another on the basis of their size, shape and orientation in space. The parameters are expressed in terms of different numbers called quantum numbers.

Quantum numbers may be defined as a set of four numbers with the help of which we can get complete information about all the electrons in an atom. It tells us the address of the electron i.e., location, energy, the type of orbital occupied and orientation of that orbital. There are four quantum numbers used to explain the properties of the electron.

As already mentioned three quantum numbers characterize the wave function and hence the orbital associated with each energy state of the hydrogen atom (or each energy value of the electron in the hydrogen atom). These quantum numbers are called (i) primary (principal) quantum number, (ii) secondary (azimuthal) quantum number, (iii) magnetic quantum number and (iv) spin quantum number.

Principal quantum number (n)

It tells the main shell in which the electron resides and the approximate distance of the electron from the nucleus. The value of n represents the shells or energy levels in which the electron is revolve around the nucleus. Different energy level in an atom have been represented by 'n' which have nonzero positive integer values up to infinity i.e., n = 1,2,3,4... Thus n can have infinite values and it has been called principal quantum number by Bohr. The maximum number of electrons that a shell can accommodate is only $2n^2$ where n is the principal quantum number. These shells or energy levels are designated by capital letters K, L, M, N...... For example the shell for which n = 1 is called K shell, n = 2 is called L shell.

Shell ————	K	L	M	N	————
n ————	1	2	3	4	————
number of electrons —	2	8	18	32	————

n value also determines the energy of an electron and also the size of the shell. More is the value of n, more far is the electron from the nucleus e.g., the electron from which n = 2 is more far from the nucleus than that for which n = 1. Similarly the electron having n = 3 will be far from the nucleus than an electron with n = 2.

Therefore :

n – values - $1 < 2 < 3 < 4 <$ ....

r – values - $r_1 < r_2 < r_3 < r_4 <$ ....

E – values - $E_1 < E_2 < E_3 < E_4$ ....

Azimuthal or Subsidiary quantum number

Azimuthal quantum number represents the various sub shells present in the main energy level. This also describes the shape or the orbital. For a given value of main energy level. 'n', 'l' can have integral values 0 to n – 1. i.e., l = 0, 1, 2, 3, 4 .... (n – 1). Thus l can have n values. Different values of l represent different sub shells which are designated by s, p, d and f ....

Value of 'l'	=	0	1	2	3	———
Designation of sub shell	=	S	P	d	f	———

The electrons with 'l' = 0, 1, 2, and 3 are said to be placed in s, p, d and f sub shells respectively. As a result we can predict the locality of electron in the sub shells. The values of l define the shape of the sub shell (electron cloud) occupied by the electron i.e., whether the sub shell is spherical dumbbell or with some other complicated shape, each value of 'l' gives a particular shape to the sub shell. For example a sub shell with l = 0 has a spherical shape, that with 'l' = 1 has a dumbbell shape, while that with 'l' = 2 has a shape with double dumbbell shape. Sub shells which have still higher values of 'l' have complicated shape.

'l' values enables us to calculate the total number of sub shells which a main shell have. Total number of sub shells in a main shell with principal quantum number 'n' is equal to the total number of 'l' values for the given value of 'n'.

For example when n = 1, 'l' = 0 since 'l' has only one value, K shell has only one sub shell that is designated as Is.

Similarly for	n = 2,	l = 0, 1.	Designation of the sub shell 2s, 2p
	n = 3,	l = 0, 1, 2.	Designation of sub shell 3s, 3p, 3d.

n = 4, l = 0, 1, 2, 3. Designation of the sub shell 4s, 4p, 4d, 4f.

'l' value is also a measure of the orbital angular momentum of an electron during its orbital motion about the nucleus. Therefore 'l' is also called angular momentum. The orbital angular momentum of the electron is given by $\sqrt{l(l+1)} \times \frac{h}{2\pi}$ for a particular value of n.

Magnetic quantum number

An electron revolving around the nucleus generates an electric field. Under the influence of external magnetic field the electrons of a sub shell can orient them selves in certain preferred regions of space around the nucleus called orbital. The magnetic quantum number determines the number of preferred orientations of the electron present in a sub shell. Different values of m for a given value of 'l' give us the total number of different ways, the sub shells are oriented in space along x, y and z axes.

For a given value of 'l', m can have integral values ranging from –l to +l through 'O'.

i.e., m = 0, ± 1, ± 2, ± 3, ± 4,....... ± l.

consequently when l = 0, m = 0

l = 1 m = -1, 0, + 1.

l = 2 m = -2, -1, 0, +1, + 2

l = 3 m = -3, -2, -1, 0 +1, +2, +3.

The spin quantum number

The emission spectrum of hydrogen and sodium can be further split by the application of an external magnetic and electric field. The only way to explain these results was to assume that electrons behave like tiny magnets. It is found that an electron in an atom not only revolves around the nucleus but also spins about its own axes. Since an electron can spin either in clock wise direction or in anti clock wise direction. Therefore, for any particular value of magnetic quantum number, spin quantum number can have two values. i.e.

and $+\frac{1}{2}$. These are represented by two arrows pointing in opposite directions, i.e., ↑ and ↓. When an electron occupy a vacant orbital, it can have clock wise spin i.e., or . This quantum number helps to explain the magnetic properties of the substances.

Quantum number and energy level relationship in atoms

Shell	n	l	Sub shell	No of sub shells	m
K	1	0	S	1	0
L	2	0	S	2	0
		1	P		-1, 0+1
M	3	0	S	3	0
		1	P		-1, 0+1
		2	d		-2, -1, 0+1, +2
N	4	0	S	4	0
		1	P		-1, 0+1
		2	d		-2, -1, 0+1, +2
		3	g		-3, -2, -1, 0+1, +2 +3

Orbital : An orbital is the region of space around the nucleus within which the probability of finding an electron of given energy is maximum. The shape of this region gives (electron cloud) the shape of the orbital. The shape of the orbital is basically determined by azimuthal quantum number 'l', while the orientation of the orbital depends on the magnetic quantum number (m).

Difference between orbit and orbital

Orbit	Orbital
1. It is a circular path around the nucleus in which the electron revolves around.	1. It is a region of space around the nucleus in which probability of finding electron is maximum
2. It represents the movement of electron in one plane.	2. It represents three dimensional space around the nucleus.
3. Its shape is circular	3. Its shape may be spherical dumbbell etc.
4. The position and velocity can be determined precisely.	4. It is impossible to find the position and velocity of the electron at any instant with certainty.

Shapes of orbitals

The shape of an orbital is determined by the value of the azimuthal quantum number ('l').

s – orbital : All s orbitals are spherical in shape but differ in size, which increases as the principal quantum number increases, there is an s orbital for each permissible value of n. It is denoted as 1s, 2s, 3s, 4s and So on. The size of the S orbital are as follows 1s < 2s < 3s < 4s ......

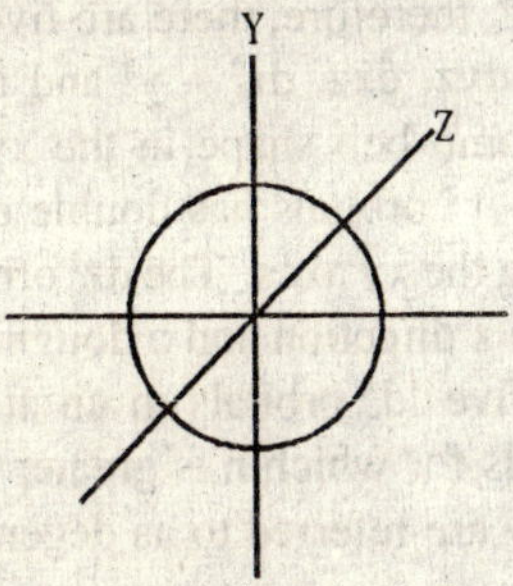

Fig 2.6: s - orbital

p – orbitals : When the azimuthal quantum number l = 1, the magnetic quantum number m can have values -1, 0, 1, therefore, we have three p orbital, each p orbital has a dumbbell shape with two lobes. The three 'p' orbitals are directed at right angles to each other with orientations along X, Y and Z axes in space. They are designated as P_x, P_y and P_z orbitals respectively.

[The letter subscripts indicate the axes along which the orbitals are oriented. These three p orbitals are identical in size, shape, and energy but they differ from one another only in their orientations]

The boundary surface diagrams for P orbitals are as shown.

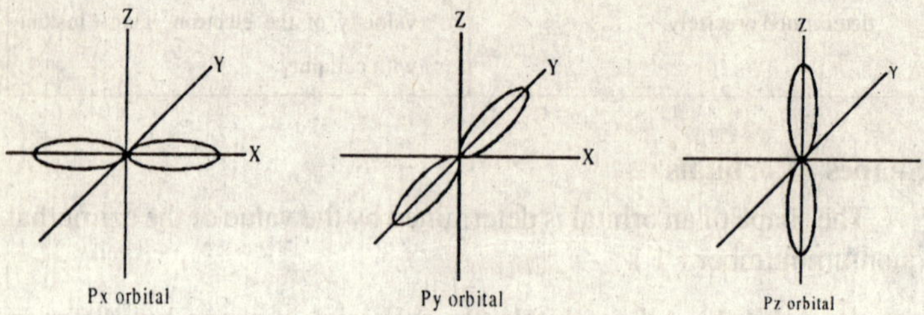

Fig. 2.7 : Shape of p - orbitals

d – orbitals

When the azimuthal quantum number l = 2, it is known as 'd' orbitals. The azimuthal quantum number l = 2 then magnetic quantum number -2, -1, 0, +1, +2, therefore, there are five 'd' orbitals. They are designated as dxy, dyz, dxz, $dx^2 - y^2$ and dz^2. The first three orbitals have double dumb bell shape in the xy, yz and xz planes respectively. The $dx^2 - y^2$ orbitals has double dumbbell shape and lies in the xy plane along the xy axes. The dz^2 orbital has a dumb bell shaped region along the z direction and a doughnut – shaped region in the xy plane. The five 'd' orbitals in an atom are identical in energy. [The 'd' orbitals for which n is greater than 3 have similar shapes. Therefore, they are referred to as degenerate orbitals (Five fold degenerate orbitals).

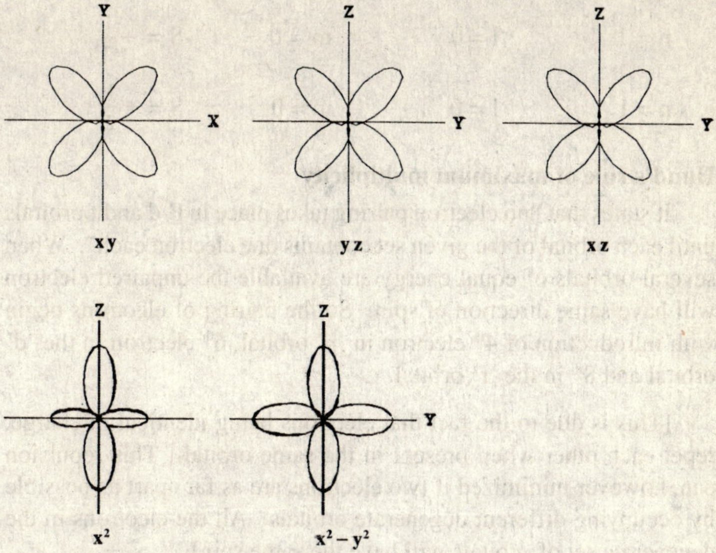

Fig. 2.8 : Shape of d - orbitals

f – orbitals: When l = 3, the orbitals are called f orbitals the shapes of these orbitals are highly complicated.

Each orbital can accommodate two electrons. Therefore, the maximum number of electrons in 's' sub shell is 2, 6 in the 'p' sub shell, 10 in the 'd' sub shell and 14 in the 'f' sub shell.

Pauli's exclusion principle

Pauli's exclusion principle states that "It is impossible for any two electrons in the same atom to have all the four quantum number same.

Therefore in the same atom, any two electrons may have three quantum numbers identical but not the fourth one. For example, consider He atom which has two electrons in the K shell, the n value is 1 and l value is 'o' for both the electron. However, spin quantum number is either $+\frac{1}{2}$ or $-\frac{1}{2}$. Thus there are two possibilities.

n = 1	l = 0	m = 0	$S = +\frac{1}{2}$
n = 1	l = 0	m = 0	$S = +\frac{1}{2}$

Hund's rule of maximum multiplicity

It states that "no electron pairing takes place in P, d and f orbitals until each orbital of the given set contains one electron each". When several orbitals of equal energy are available the unpaired electron will have same direction of spin. So the pairing of electrons begin with introduction of 4th electron in 'p' orbital, 6th electron in the 'd' orbital and 8th in the 'f' orbital.

[This is due to the fact that electrons being identical in charge, repel each other when present in the same orbital. This repulsion can, however minimized if two electrons are as far apart as possible by occupying different degenerate orbitals. All the electrons in the degenerate set of orbitals will have the same spin].

Illustration : Let us consider the distribution of three electrons in three degenerate orbitals, namely p_x, p_y and p_z, there are many different ways of arranging the three electrons in p_x, p_y and p_z orbitals. Only three of these arrangements are shown below.

p_x	p_y	p_z		p_x	p_y	p_z		p_x	p_y	p_z
↑↓	↑			↑	↓	↑		↑	↑	↑

Arrangement (a) Arrangement (b) Arrangement (c)

According to Hund's rule of maximum multiplicity arrangement 'C' is more stable of all the three arrangement. Since it contain maximum number of unpaired electrons having parallel spins.

Illustration : Consider a nitrogen atom which has seven electron. According to Hund's rule, the electron distribution as follows.

N (z = 7)	↑↓	↑↓	↑	↑	↑
	$1s^2$	$2s^2$		$2p^3$	
O (z = 8)	↑↓	↑↓	↓↑	↑	↑
	$1s^2$	$2s^2$		$2p^4$	
Ne (z = 10)			↑↓	↑↓	↑↓
	$1s^2$	$2s^2$		$2p^6$	

Electronic configuration

The term electronic configuration of an atom tells as about how the electrons are distributed among various atomic orbitals. The electronic configuration of an atom can be represented as

$$nS^x$$

x = Denotes the number of electrons in the orbital or sub shell.

S = Denotes the angular momentum number or azimuthal quantum number (l)

n = Denotes the principal quantum number

Ground state electronic configuration

Atomic number	*Symbol*	*Electronic Configuration*
1	H	$1S^1$
2	He	$1S^2$
3	Li	$1S^2 2S^1$
4	Be	$1S^2 2S^2$
5	B	[He] $2S^2 2P^1$
6	C	[He] $2S^2 2P^2$
7	N	[He] $2S^2 2P^3$

8	O	[He] $2S^2\ 2P^4$
9	F	[He] $2S^2\ 2P^5$
10	Ne	[He] $2S^2\ 2P^6$
11	Na	[Ne] $3S^1$
12	Mg	[Ne] $3S^2$
13	Al	[Ne] $3S^2\ 3P^1$
14	Si	[Ne] $3S^2\ 3P^2$
15	P	[Ne] $3S^2\ 3P^3$
16	S	[Ne] $3S^2\ 3P^4$
17	Cl	[Ne] $3S^2\ 3P^5$
18	Ar	[Ne] $3S^2\ 3P^6$
19	K	[Ar] $4S^1$
20	Ca	[Ar] $4S^2$
21	Sc	[Ar] $4S^2\ 3d^1$
22	Ti	[Ar] $4S^2\ 3d^2$
23	V	[Ar] $4S^2\ 3d^3$
24	Cr	[Ar] $4S^1\ 3d^5$
25	Mn	[Ar] $3d^5\ 4S^2$
26	Fe	[Ar] $4S^2\ 3d^6$
27	Co	[Ar] $4S^2\ 3d^7$
28	Ni	[Ar] $4S^2\ 3d^8$
29	Cu	[Ar] $4S^1\ 3d^{10}$
30	Zn	[Ar] $4S^2\ 3d^{10}$
31	Ga	[Ar] $4S^2\ 3d^{10}\ 4P^1$
32	Ge	[Ar] $4S^2\ 3d^{10}\ 4P^2$
33	As	[Ar] $4S^2\ 3d^{10}\ 4P^3$

34	Se	[Ar] $4S^2$ $3d^{10}$ $4P^4$
35	Br	[Ar] $4S^2$ $3d^{10}$ $4P^5$
36	Kr	[Ar] $4S^2$ $3d^{10}$ $4P^6$
37	Rb	[Kr] $5S^1$
38	Sr	[Kr] $5S^2$
39	Y	[Kr] $5S^2$ $4d^1$
40	Zr	[Kr] $5S^2$ $4d^2$
41	Nb	[Kr] $5S^1$ $4d^4$
42	Mo	[Kr] $5S^2$ $4d^5$
43	Te	[Kr] $5S^2$ $4d^5$
44	Ru	[Kr] $5S^2$ $4d^7$
45	Rh	[Kr] $5S^2$ $4d^8$

General configuration of S, P, and d and block elements

S – block elements – $nS^{1\text{-}2}$

P – block elements – $nS^{1\text{-}2}$ $nP^{1\text{-}6}$

d – block elements – $(n-1)\ d^{1\text{-}10}$ $nS^{1\text{-}2}$

QUESTIONS CARRYING ONE MARK

1) Name the three fundamental particles of an atom.
2) Name the scientist who discovered the electron.
3) How many electrons s and p sub shell can accommodate?
4) How many protons and neutrons are present in Cl?
5) What are cathode rays composed of ? What are anode rays composed of?
6) Neutrons are present in all atoms except ________ atom.
7) How is an orbital with n = 4 and l = 3 is designated?
8) Give the values of l and n for 3p sub shell.
9) Write the formula that gives maximum number of electrons in a sub shell.

10) Write the possible value of l when n = 3.
11) When n = 1, indicate the total number of possible orbitals ?
12) What experiment suggest that the electron is a universal constituents of matter ?
13) What is the name of the stream of electrons flowing from the cathode towards the anode?
14) What is an orbital?
15) Who discovered neutrons?
16) Define atomic number of an electron?
17) What is the absolute charge of an electron?
18) Define mass number of an element.
19) Define quantum number.
20) What are the value of azimuthal quantum number when n = 2 ?
21) What is the shape of the orbital when l= 2 ?
22) Maximum number of electrons that can be accommodated in an orbital is ____.
23) Neutrons and protons are collectively called ____.
24) Write the Rydberg's formula for the wavelength of first Balmer line in Balmer series.
25) What is (n + l) rule ?
26) How many unpaired electrons are present in chromium atom (At. no. 24) in it's ground state ?
27) Arrange the following sub shells in the increasing order of energy: 4s, 4p, 4d, 4f, 5s, 5p, 5d and 5f.
28) What is the atomic number of element having the electronic configuration $3d^{10}\ 4s^1$?
29) State Pauli's exclusion principle.
30) State Heisenberg's uncertainty principle.
31) Wavelength range of visible light is ____.
32) Who proposed quantum theory ?
33) What is the value of Planck's constant ?
34) What is the shape of orbital for $l = 0$ and $l = 1$?
35) The average distance of the electron from the nucleus in an atom is indicated by which quantum number ?

36) Magnetic quantum number describes ______.

37) What possible value are there for spin quantum number ?

38) How many unpaired electrons are there in a Al^{+3} ion ?

39) The number of degenerate orbitals in the f sub shell are _______.

40) Write the electronic configuration of Cr atom.

41) Write the electronic configuration of Mn^{4+} ion.

42) How many electrons are there in the valence shell of aluminium atom ?

43) Elements of same Atomic number but of different mass number are known as ______.

44) The $2p_x$, $2p_y$ and $2p_z$ orbitals of an atom have identical shape but different in their _______.

45) State Hunds rule of maximum multiplicity? Give an example.

QUESTIONS CARRYING TWO MARK

1) What are the difference between an orbit and an orbital ?

2) Give reason "there are many lines in the hydrogen spectrum even though hydrogen atom contains only one electron".

3) Write the Rydberg formula for the wavelength of spectral line in Balmer series.

4) Write de Broglie equation and explain the terms.

5) What is meant by atomic spectra? How are Lyman lines formed in hydrogen spectrum ?

6) Sketch the shapes of s and p orbitals indicating orientation of p - orbitals.

7) How many orbitals are there in M - shell? How are they designated ?

8) State Aufbau principle. Illustrate with an example.

9) Give reason "the outer electronic configuration of copper is $3d^{10} 4s^1$ and not $3d^9 4s^2$".

10) How are the five - d orbitals designated ?

11) Nitrogen has 3 unpaired electron. Explain using Hund's rule of maximum multiplicity.

12) Write the number of unpaired electrons in the atoms of the following in the ground state. Carbon, Nitrogen, Chlorine, and Sulphur.
13) Write the atomic numbers of the elements with the outer electronic configuration.

    (a) $3s^2$ (b) $4s^2 3d^3$
14) Write the "n" and "*l* " value for the following orbitals.

    (a) 2s (b) 5p (c) 3d
15) Write the magnetic quantum numbers of the electrons in the seven orbitals of 4f level.
16) Write the electronic configuration of Fe^{2+} and Fe^{3+} ions. Which is more stable?
17) Give reason: half - filled and completely filled orbitals are highly stable.
18) Apply (n + *l*) rule and show that 4s level is of lower energy than 3d level.
19) Distinguish between frequency and wave number.
20) Give reason : different gases produce the same cathode rays but different positive rays.
21) Distinguish between atomic number and mass number.
22) Write a note on discovery of proton.
23) How many unpaired electrons are there in Ni^{+2} ion?
24) When n = 3, indicate the total number of possible orbitals?
25) Write a note on cathode rays.
26) Sketch the shape of p orbitals.
27) Distinguish between shell and sub - shell.
28) Write all the possible magnetic quantum numbers when l= 3.
29) 'p' orbitals are three - fold degenerate and 'd' orbitals are five - fold degenerate. Explain.
30) Which series of hydrogen spectrum lies in the visible region and UV region?

QUESTIONS CARRYING FOUR MARK

1) Explain with an example Hund's rule of maximum multiplicity.
2) What is the symbol for the subsidiary quantum number ? What are its values ? What property of the electron wave is associated with this quantum number ?
3) What is the symbol for the spin quantum number ? Why was it introduced ? What are its values ?
4) Draw the shapes of 3d orbitals.
5) Draw the shapes of 2s and 2p orbitals.
6) Give the electronic configuration of the following ions.
   (a) H^- (b) Na^+ (c) F^- (d) Mg^{2+}
7) What is (n + I) rule? Explain with example.
8) Write that electronic configuration of elements with atomic number 24, 25 and 29.
9) Derive an equation which is used for explaining wave particle nature of electron.
10) Explain Debroglie's wave - particle theory with illustration.
11) Give an account of emission spectrum of hydrogen and explain the existence of various series in the spectrum.
12) What are Quantum numbers? Discuss the significance of the four Quantum numbers of an atom.
13) Give an account of the discovery of
    (a) Cathode Rays (b) Proton (c) Neutron
14) State the essential postulates of Bohr's model of the atom. How does it explain the spectrum of hydrogen?
15) Discuss the merits and demerits of Bohr's model of the atom.
16) Explain the terms wavelength, frequency and velocity of light.
17) (a) State any four postulates of Bohr's theory.
    (b) Write the possible values for *l* and m, for an electron in 3d orbital.
    (c) Write the electronic configuration for K and Zn.
    (d) State Pauli's exclusion principle.

18) (a) What are quantum numbers? Explain any two quantum numbers.

(b) State Hund's rule of maximum multiplicity.

(c) Draw the shapes of all 'd' orbitals.

(d) Which designation is given to the orbital having

(i) $n = 2 \quad l = 1$ (ii) $n = 1$ and $l = 0$

19) (a) Explain, how the emission spectrum of hydrogen is obtained.

(b) Demerits of Bohr's theory.

(c) What do you understand by the term dual nature of matter.

(d) Explain, why two electrons of an atom Cannot have same *n, m, l* and *s* values.

20) Calculate the wavelength of a wave of frequency 10^{12} Hz, traveling with the speed of light.

Ans: (l = 3 x 10^{-4} m)

21) Calculate the frequency of electro magnetic radiation having the wavelength 500 nm. Also calculate the wave number corresponding to it.

Ans: (ν = 6 x 10^{14} Hz ; = 2 x 10^{6} m^{-1})

22) Calculate the frequency and energy per quantum of a radiation with a wavelength of 300nm. (c = 3 x 10^{8} ms^{-1} and h = 6.625 x 10^{-34} Js) (ν = 10^{15} s^{-1} and E = 6.625 x 10^{-19}J)

23) A major line in an atomic spectrum occurs at 450 run. Find the energy decrease as this photon is emitted.

Ans: (4.417 x 10^{-19} J)

24) Calculate the wave number, wavelength and frequency of the first line in the Balmer series. R_H = 10967800m^{-1}.

Ans: (= 1.523 x 10^{6} m^{-1} ; ν = 4.567 x 10^{14} s^{-1} ; λ = 656.5 nm)

25) The red light of Neon has a wavelength of 693 nm. Find the energy difference (per mole of atoms) between the two energy levels involved.

Ans: (Δ E = 172.7 KJ)

26) Calculate the wavelength of an electron moving with a velocity 2.5×10^{8} ms^{-1}. $h = 6.625 \times 10^{-34}$ Js. Mass of an electron = 9.11×10^{-31} kg.

Ans: ($\lambda = 2.9 \times 10^{-12}$m)

27) Find the mass of an electrically charged particle moving with a velocity of 3×10^{6} ms^{-1} and having the de Broglie wavelength of 2.

Ans : (Mass = 1.1042×10^{-30}kg)

28) Calculate the energy of a photon of frequency, 4×10^{15} s^{-1}.

Ans : (E=2.65×10^{-18}J)

29) Calculate the wavelength of a bullet of mass 2g moving with velocity of $3.313 \times 10^{4} \times ms^{-1}$.

Ans : ($\lambda = 1 \times 10^{-36}$ m)

30) The uncertainty in the momentum of a particle is 1×10^{-10} kg ms^{-1}. Find the accuracy with which position of the particle can be determined?

Ans : ($\Delta x = 0.527 \times 10^{-24}$ m)

31) Calculate the wave number and energy of the spectral line of hydrogen corresponding to $n_1 = 3$ and $n_2 = 5$. The Rydberg constant $R = 1.097 \times 10^{7}$ m^{-1}

Ans : 7.8×10^{5} m^{-1}.

3

PERIODIC TABLE

Introduction

This is a table in which elements are arranged in the order of increasing atomic number in the manner that the elements with similar properties fall in the same vertical column. The elements which fall in the same column and resemble one another in their properties are said to belong to the same group or family of elements.

In 1869 Mendeleef enunciated periodic law as "the physical and chemical properties of elements are periodic functions of their atomic weights" However, there was a drawback in this form of periodic table

In 1902, Moseley determined the atomic number of elements. He found that atomic number is more fundamental property than atomic weight. He stated the modern periodic law. The long form of periodic table based on Moseley is modern periodic law. It states that "The properties of elements are a periodic functions of their atomic numbers".

The long form of periodic table constituted by arranging elements in increasing order of their atomic number. The horizontal rows constitutes periods and vertical rows are called groups.

There are seven horizontal rows, called periods and 18 vertical columns of elements, these are called groups. They are numbered 1 to 18 according to IUPAC.

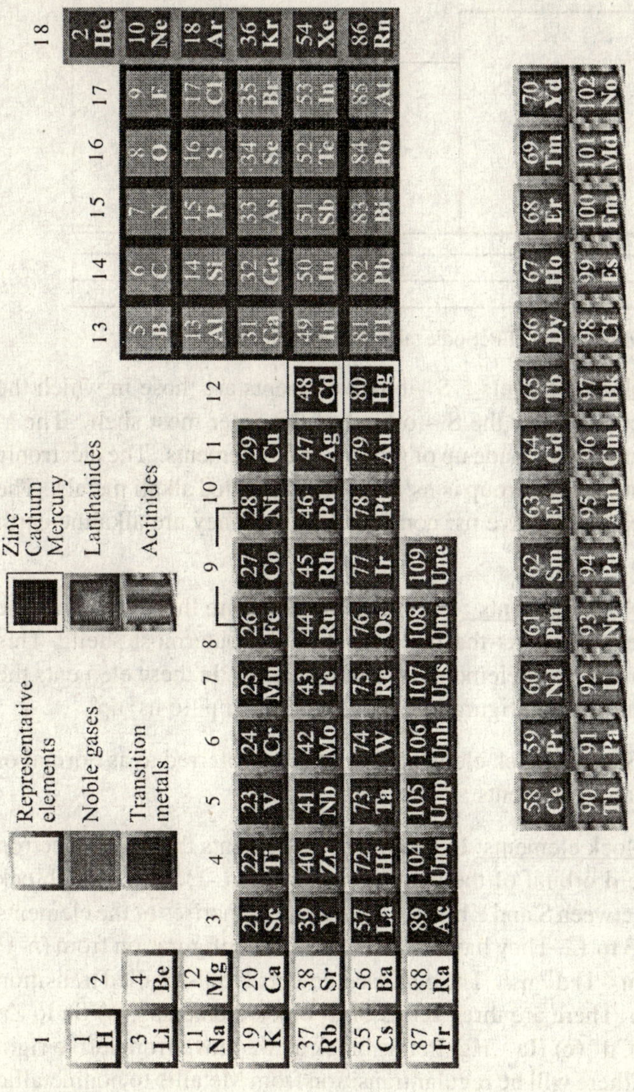

Fig : 3.1 Modern periodic table

S. P. d. and f blocks: On the basis of electronic configuration, the long form of the periodic table can be divided into four blocks, they are s. p. d. and f blocks.

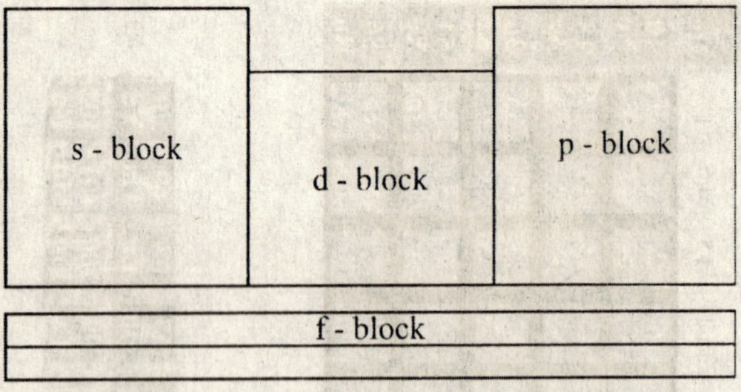

Fig : 3.2 Arrangement of Periodic table into four blocks.

S – block elements : S – block elements are those in which the added electron enter the S – orbital of the outer most shell. The s-block elements are made up of group 1 and 2 elements. The electronic configuration of I^{st} group is ns^1 and they are called alkali metals. The group 2 elements have ns^2 configuration and they are alkaline earth metals.

P – block elements : P – block elements are those in which the added electron enters the p – orbital of the outermost shell. This block is made up of elements of group 13 to 18. In these elements the outer electronic configuration varies from $ns^2\ np^1$ to $ns^2\ np^6$

The S and P block elements are together referred to as normal or representative elements.

d – block elements: In this block of elements the added electron enters the d orbital of the penultimate shell (n -1). Thus, d – block appears between S and P blocks. This block comprises of the elements of group 3 to 12. They have outer electronic configuration from (n-1) $d^1\ ns^2$ to $(n-1)\ d^{10}\ ns^2$. The d block elements are also called transition elements. There are three series of d- block elements: **(a)** Sc to Zn **(b)** Yt to Cd **(c)** Ha – Hg. In this block as we move from left to right of period there will be regular transition from Metallic to nonmetallic character f – block elements: In these elements, the added electrons occupies f – orbitals of the antepenultimate shell, (n -2) there are two sets of f – block elements. They are 4f series (lanthanides) and 5 f series (actinides). There are 14 elements in 4 f series and 14 elements in 5f series. Lanthanides and actinides together are called inner transition elements.

Cause of periodicity: The term periodicity refers to the recurring of properties of elements at a certain regular intervals of atomic number in the periodic table. The cause of periodicity is due to the recurrence of similar electronic configurations in the valence shell of elements

According to the modern periodic law, the properties of elements like atomic radius, ionic radius, Ionization energy, electron affinity, electronegativity etc are the periodic functions of their atomic numbers. This means, with the increase in atomic number of the elements in the same period or group there is a gradual variation in a particular property of the element.

Description of Periodic table

Periods The horizontal rows are called periods. There are seven periods in the periodic table.

I Period: The first period is made up of only two elements namely hydrogen and helium. The principal quantum number is n = 1

II Period: The second period contains 8 elements. It starts with lithium and end with neon. The principal quantum number is n = 2.

III Period: The third period contains 8 elements. It starts with alkali metal sodium and ends with noble gas argon. The principal quantum number is 3 for this period. They have three main shells i.e., K, L, and M shells. The second and the third periods are short periods.

IV Period: This period consists of 18 elements from potassium (Z = 19) to the noble gas Krypton (Z = 36). Potassium is an alkali metal and calcium is an alkaline earth metal. There are 10 d – block elements i.e., from Scandium to zinc. Scandium has an outer electronic configuration $(n - 1)d^1 ns^2$ and zinc has $(n - 1)d^{10} ns^2$.

There are six p – block elements are Gallium (Z = 31) to Krypton (Z = 36). The outer electronic configuration of Gallium is ns^2 np^1 and krypton is ns^2 np^6. The elements, potassium and calcium and Gallium to krypton are called normal or representative elements. The difference between the normal and transition elements lies in the manner in which the electrons are added to valence shell. Transition metals differ from normal elements in many respects like showing variable valency, forming coloured ions and being paramagnetic.

<u>V Period:</u> Fifth period also contains 18 elements. It begins with Rubidium (Z = 37) and ends with Xenon (Z = 54) out of 18 elements, eight are normal elements and ten are transition elements.

<u>VI Period:</u> This long period consists of 32 elements they include 8 normal elements (Z = 55, 56 and 81 to 86), 10 transition elements (Z = 57 and 72 to 80) and 14 rare earth elements (Z = 58 to 71).

The rare earth element start with cerium (Z = 58) ends with Lutetium (Z = 71). The rare earth elements are similar to lanthanum, they are collectively called Lanthanides. If these elements are placed in the horizontal rows, these will be under expansion of the table. To avoid this, they are placed as a separate series below the periodic table.

<u>VII Period:</u> The seventh period consist of 20 elements at present. (Z = 87 to Z = 106). These are called actinides because they are all similar to actinium. These elements are also kept below the periodic table separately. The elements beyond uranium have been prepared by artificial transmutation are called transuranium elements.

Groups

The elements of group one and two are called alkali metals and alkaline earth metals respectively. The elements of group 13 to 17 are p – block elements. The halogen in group 17 show most complete set of properties of nonmetals. Group 18 elements are called noble gases or inert elements because they possess complete octet structure. They are chemically inert and majority of them exist mono atomic gases.

Periodic properties of elements in modern periodic table

Periodic variation in physical properties

It has been observed that, the electronic configuration of elements show a periodic variation with increase in atomic number. Consequently, the elements also display periodic variations in their physical and chemical behaviour.

1. Atomi radius

Atomic radius is defined as one half the distance between the nuclei in two adjacent metal atoms.

For elements that exist as simple diatomic molecules, the atomic radius is one half the distance between the nuclei of the two atoms in a particular molecule. The internuclear distance can be obtained by x-ray and spectroscopic methods.

Atomic radius depends on the nature of binding which in turn influences the electron could. There are two types of atomic radii are recognized. (i) covalent radius (ii) Vander waals radius.

Covalent radius: Half of the bond length of a covalent bond between two like atoms. Example molecule A – A.

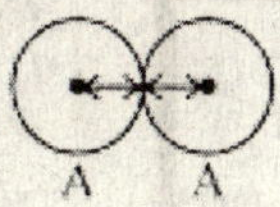

Atomic radius

The single bond distance between carbon – carbon in diamond is found to be 154 pm. So, the covalent radius of carbon $= \frac{154}{2} = 77 \text{Pm}$

Vander waals radius: It is one half the distance between the nucleus of two adjacent atoms belonging to two neighbouring molecules. Two atoms cannot come nearer than this distance without bond formation.

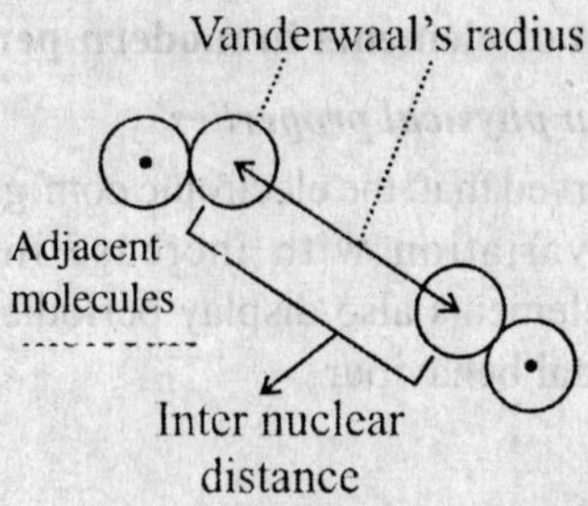

In case of homonuclear diatomic molecules of A_2 types the bond length, $d_{(A-A)}$ is given by $d_{(A-A)} = r_A + r_A = 2r_A$

$$\therefore r_A = \frac{d_{(A-A)}}{2}$$

In case of hetero nuclear diatomic molecules of A –B type. The bond length d_{A-B} is equal to the sum of the covalent radii of A and B atoms.

$d_{A-B} = r_A + r_B$

The magnitude of r_A can be calculated if d_{A-B} and r_B are known.

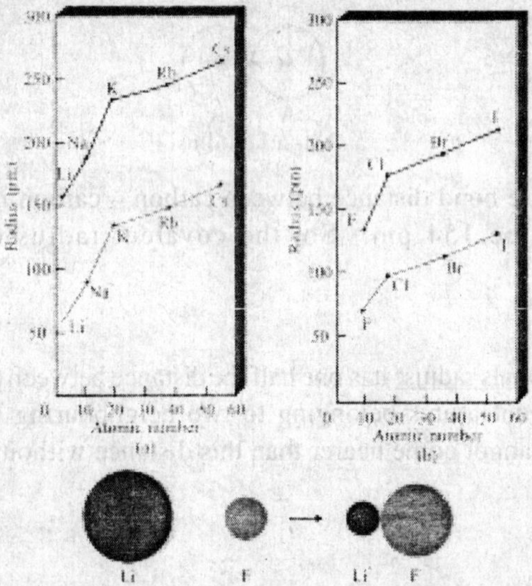

Fig : 3.3 Variation of atomic radius with increase in atomic number

Variation of atomic radii in periodic table

In group: As we go down a group, say group I elements, we find that atomic radius increases with increasing atomic number. For alkali metals, the outer most electron resides in the ns^1 orbital. Since the orbital size increases with increase in principal quantum number n, the size of the metal atom increase from Li – Cs. The same reasoning can be applied to other groups also.

In Period: Atomic radius progressively decreases from left to right of a period. This is due to increase in effective nuclear charge while the electrons are being added to the same shell. With the increase in the effective nuclear charge, the outer electrons are pulled towards the nucleus more strongly. This results in decrease of atomic size and hence atomic radius.

Ionic radius: Ionic radius is defined as "The distance between the nucleus of an ion and the point upto which the nucleus has influence on its electron cloud".

The ionic radii has additive character. This means if the ions of an ionic crystal are regarded as spheres. The internuclear distance between the two ions is equal to the sum of the radii of the ions. Thus in an ionic crystal of type $A^+ B^-$

$$d(A^+ B^-) = r_{(A+)} + r_{(B-)}$$

If the radius of one ion is known then the radius of the other ion can be calculated.

[Note: ionic radius cannot be taken as one half of the bond length of an ionic compound. This is due to the fact that size of the anion and cation is not same]

Variation in periodic table

In a period: Ionic radii decreases as we move from left to right in a period. This is due to increase in effective nuclear charge of the ion.

In a Group Within a group ionic radii increase with increase in atomic number. As we move down the group, for every element a new principal quantum shell is added. This results in increase of screening effect.

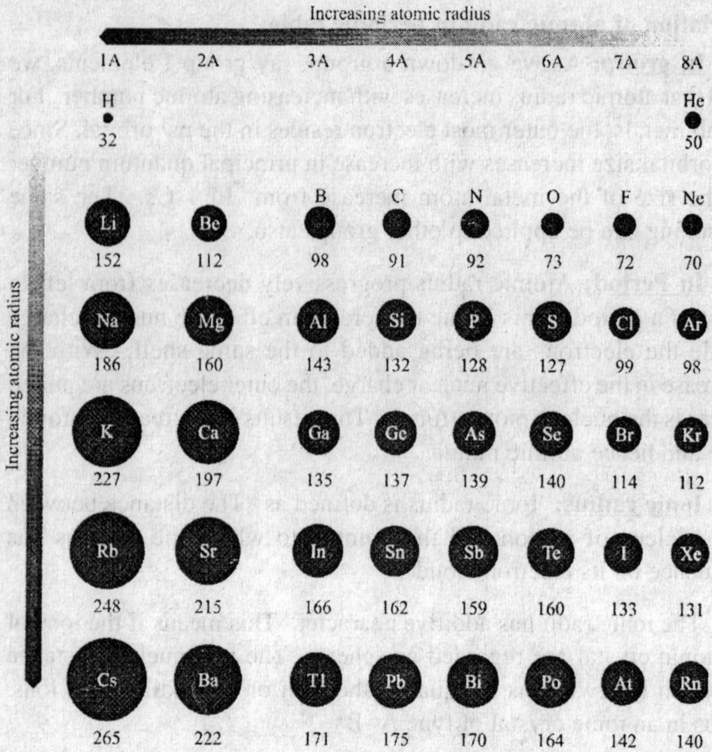

Fig : 3.4 Variation of atomic radius in periodic table

A Cation is smaller than the parent atom

The radius of an atom is altered during loss or gain of electrons by an atom. When an atom loses an electron to form a cation, the number of electrons decreases while the nuclear charge remains same. The remaining electrons are drawn more towards the nucleus. As a result, the cation has a smaller size than the corresponding neutral atom

Example: Li – 134 Li^+ – 76

Na – 154 Na^+ – 102

An anion is bigger than the parent atom

When an atom gains an electron it is converted to a negative ion. Thus an onion has more electrons than its parent atom. With the increase of electrons the shielding effect increases resulting in the decrease of effective nuclear charge. The electron – electron repulsion, over comes the attractive force of the nucleus and electrons. Therefore an anion is larger than the corresponding neutral atom.

Example: O – 74 pm, O^{2-} – 140 pm
S – 10.3 pm S^{2-} – 184 pm

Isoelectronic ions

Ions which have the same number of electrons are called isoelectronic ions. For example c^{4-}, N^{3-}, O^{2-}, F^{-}, Na^{+}, Mg^{+2} are isoelectronic ions.

ion	N^{3-}	O^{2-}	F^{-}	Na^{+}	Mg^{+2}
Number of electrons	10	10	10	10	10
Atomic number	7	8	9	11	12
Radius(pm)	171	140	136	102	72

The size of the isoelectronic ions decreases with increase in the charge on the ion. Thus the size is $mg^{2+} < F^{-} < O^{2-} < N^{3-}$

Ionization energy

Ionization energy is the minimum energy required to remove an electron from a isolated gaseous atom in its ground state.

The magnitude of ionization energy is a measure of the energy required to force an atom to give up an electron, or how tightly the electron is held in the atom. The higher the Ionization energy more difficult it is to remove an electron.

$$M(g) + IE, \rightarrow M^{+}(g) + e^{-}$$

Ionization energy is expressed in KJ mol^{-1}.

Ionization energy is also called ionization potential, since it represents the amount of potential required to remove the most loosely bound electron from the gaseous atom.

__Successive ionization energies__: After the first electron has been removed from a neutral atom, it is possible to remove second, third or even more electrons. The energies required for the removal of subsequent electrons are known as second ionization energy, third ionization energy, etc

$M (g) + IE_1 \rightarrow M^+ (g) + e$ First ionization energy.

$M^+ (g) + IE_2 \rightarrow M^{++} (g) + e$ Second ionization energy.

$M^{++} (g) + IE_3 \rightarrow M^{+++} (g) + e$ Third ionization energy.

It is observed that successive ionization energies go on increasing progressively because it becomes increasingly difficult to remove each successive electron. Therefore, $I.E_3 > I.E_2 > I.E_1$.

Variation in periodic table

In period: Ionization energy increases with the increasing atomic number as we move from left to right along a period. This is because of the following reasons.

(i) The nuclear charge increases from left to right

(ii) The number of main energy levels remain the same while the atomic size decreases. Due to the gradual increase in the nuclear charge and simultaneous decrease in the atomic size, the electron gets more and more firmly bound to the nucleus. Therefore, it becomes increasingly difficult to remove an electron and hence the ionization energy keeps an increasing.

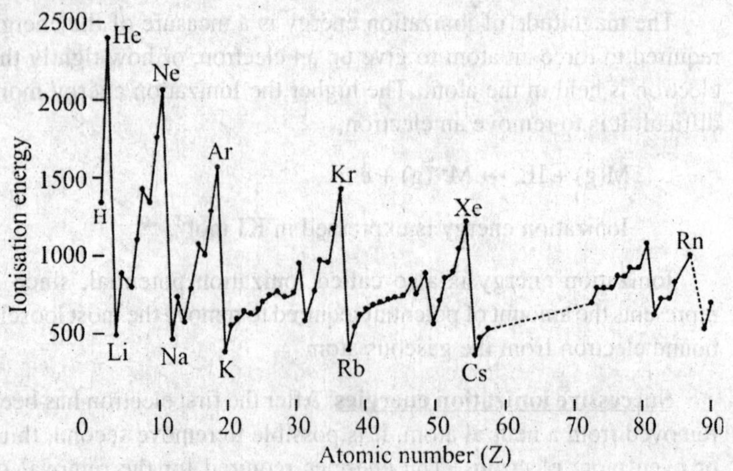

Fig : 3.5 Trends in Ionization energy with atomic number

1) Noble gases appears at peaks of the curves. They have very high ionization energy.

2) Alkali metals appears at the bottom of the curve. They have very low ionization energy. This Means it is easy to remove an electron from an atom of an alkali metal

Elements of IInd period	Li	Be	B	C	N	O	F	Ne
Nuclear charge	+3	+4	+5	+6	+7	+8	+9	+10
Ionization energy in KJ	520	899	801	1086	1402	1314	1681	2080

In group: The ionization energy decreases down the group. This can be attributed to the following factors:

a) On moving from top to bottom in a group the nuclear charge increases.

b) There is a simultaneous increase in the number of energy levels and therefore, the atomic size increases.

c) There is an increase in the number of electrons that lie between nucleus and the outermost electron and therefore, the shielding effect in the outermost electron gradually increases.

The increase in atomic size and the shielding effect more than neutralize the effect of the increased charge. As such the electron in the higher orbital is less firmly bound to the nucleus. Consequently there is gradual decrease in the ionization energy.

Electronegativity

Electronegativity is defined as power of an atom in a molecule to attract electrons to it self. Greater the ability of an atom to attract electron, larger is the value of electronegativity. This is different from electron affinity since it is concerned with atoms in a molecule and not with isolated atoms

In homonuclear diatomic molecules like O_2, N_2, Cl_2 etc the electron pair is.equally shared between the atoms because both the atoms have equal attraction towards the bonded pair of electrons, These atoms have equal electronegativity. Such molecules are called <u>nonpolar</u>.

In heteronuclear diatomic molecules like HCl, HF etc., the bonded pair is not equally shared and the atom which is more electronegative has more attraction towards the electron pair, The more electronegative atom acquires a slight negative charge and the other atom gets slight positive charge. Thus there will be charge separation between bonded atoms and the molecule will be polar.

In Polyatomic molecules, the polar and nonpolar nature depends on electronegativity values of each atom and arrangement of atoms in the molecule. The molecules with symmetrical structures like CCl_4, CO_2 are non polar and the molecules with asymmetrical structures like H_2O and NH_3 are polar in nature.

Variation in periodic table:

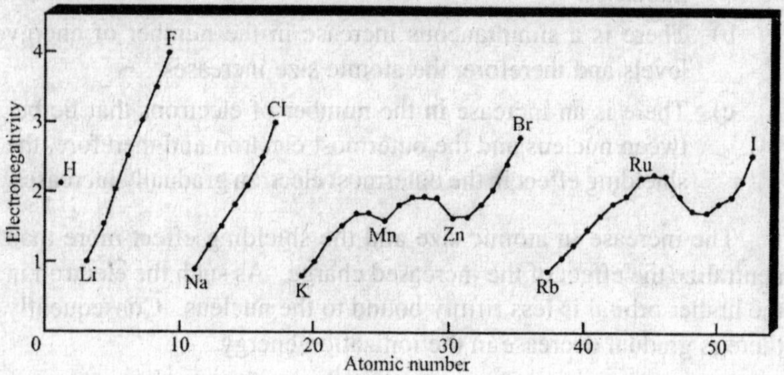

Fig : 3.6 Relationship between electronegativity and atomic number

The electronegativity depends on many factors. They are (i) the charge on the nucleus (ii) the distance of the outer electrons from the nucleus (iii) Shielding effect of the inner electrons.

In a period: The electronegativity of elements increases from left to right in a period.

In a group: The electronegativity decreases from top to bottom of a group.

Pauling proposed the scale of electronegativity. He assigned a value of 4.0 for most electronegative element fluorine. According to

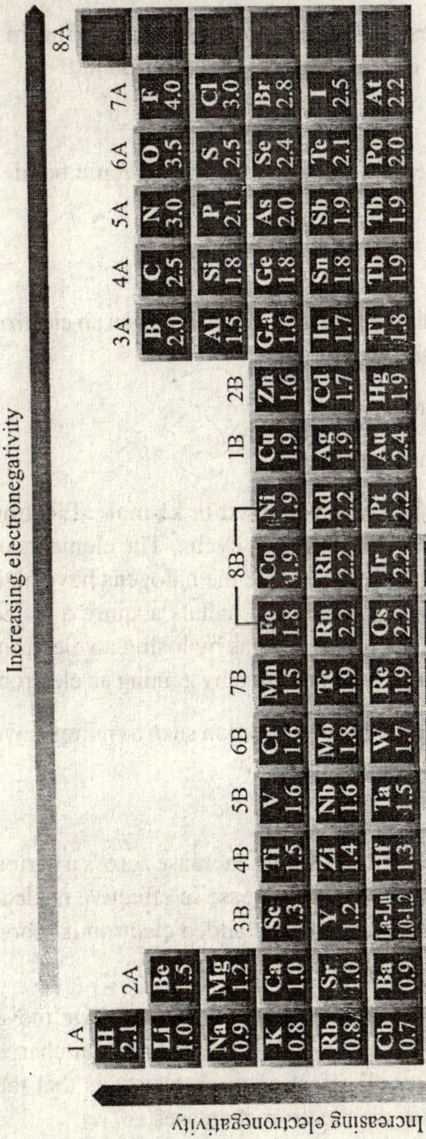

Fig : 3.7 Relationship between electronegativity and atomic number in periodic table

him caesium has 0.7. The value of other elements lie in between these two values. Noble gases are not given any electronegativity values.

Electronegativity values can predict the type of bonding in a molecule. If the difference is zero then it is nonpolar bond .

Ex:- H – H Molecule.

If the difference is large than the bond formed, is ionic bond

Ex:- SO_2 , H_2O.

Electron affinity

Electron affinity is the energy change that occurs when an electron is accepted by an atom in the gaseous state.

X (g) + e X^- + energy released.

Where x is an atom of an element.

The electron affinity values are expressed in kJ mol^{-1}. Electron affinity values are calculated by Born Haber cycle. The elements of first group have low electron affinities while the halogens have high electron affinity. This is because the alkali metals acquire a stable configuration (ns^2np^6) of the nearest noble gas by losing an electron. The halogen atoms acquire stable configuration by gaining an electron.

Elements with stable electronic configuration such as $ns^2 np^6$ have positive values for electron affinity.

Trends in periodic table

<u>**In a period:**</u> The electron affinity values increase across a period due to decrease in the size of atom with increase in effective nuclear charge. The attraction between nucleus and added electron is more, as a result, the formation of negative ion becomes easy.

<u>**In a group:**</u> The electron affinity values decrease as we move down a group. Atomic size increases with increased nuclear charge and atomic number. The over all effect of these factors is that the effective nuclear attraction for the electron decreases.

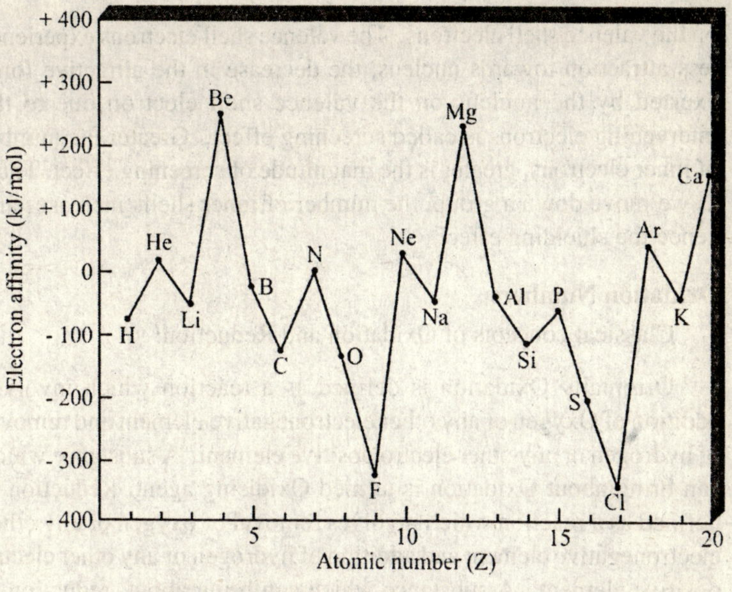

Fig : 3.8 Variation of Electron affinity with atomic number

Some important features of electron affinity values of elements:

1. The electron affinity of chlorine is more than fluorine. This is due to smaller size of fluorine than chlorine. In fluorine second shell is the outer most shell. Due to the small volume of this shell, the addition of extra electron results in strong electron - electron repulsions. As a result, attractive interaction between added electron and nucleus is less
2. The halogens have high electron affinity. This is due to the fact that addition of an electron leads to a stable electronic configuration ns^2 np^6 of the valence shell.
3. Noble gases have zero electron affinity values, since they have ns and np orbitals which are completely filled with electrons.

Shielding effect or Screening effect

In an atom containing two or more shells, the electrons present in the valence shell are attracted by the nucleus and also at the same time repelled by the electrons present in the inner shells. The combined effect of these two forces is the actual force exerted by the nucleus on the valence shell electrons. The valence shell electrons experience less attraction towards nucleus, the decrease in the attractive force exerted by the nucleus on the valence shell electron due to the intervening electrons is called screening effect. Greater the number of inner electrons, greater is the magnitude of screening effect. Thus, as we move down a group, the number of inner shells increases and hence the shielding effect.

Oxidation Numbers

Classical concepts of Oxidation and Reduction:

Originally Oxidation is defined as a reaction which involves addition of Oxygen or any other electronegative element and removal of hydrogen or any other electro positive element. A substance which can bring about Oxidation is termed Oxidising agent. Reduction is defined as a reaction which involves removal of oxygen or any other electronegative element and addition of hydrogen or any other electro positive element. A substance which can bring about reduction is termed a reducing agent.

Modern concepts of Oxidation and Reduction:

Such concept of electronic interpretation and concept of oxidation number give a broader meaning of Oxidation and Reduction

1.Electronic interpretation of Oxidation and Reduction:

(is applicable to reactions involving ionic reactivates.)

According to electronic interpretation

Oxidation is defined as reaction which involves loss of electrons by an atom, group of atoms or an ion. The substance which loser electrons is said to be Oxidised and is called a reducing agent. **Eg.:** $Na \rightarrow Na^+ + e^-$. Sodium is Oxidised to Na^+, It is a Reduction agent.

$$Na^+ OH^- + H^+ Cl^- \rightarrow Na^+ Cl^- + H_2O.$$

No state of Oxidation or Reduction.

2. OXIDATION NUMBER CONCEPT:

It is applicable to reactions involving ionic as well as nonionic reactants. This concept gives a broader meaning of Oxidation and Reduction. It does not involve transfer of Electrons.

In an ionic compound, oxidation number or oxidation state of an element is defined as the charge carried by the ion of the element in the compound. Eg.: NaCl consists of Na^+ and Cl^- ions. Oxidation number of Na is + 1 that of Cl is -1 in NaCl

In a covalent compound, Oxidation number of an element is defined as the charge which the atom of the element appears to have gained or lost when the bonded pairs of electrons (shared pairs) between two unlike atoms are counted with the more electro negative atom.

Eg.:H_2O

```
  . .
H : O :
  . .
  H
```

O is more electronegative than H. If the bonded pairs of electrons are counted with more electro negative O atom, Oxygen atom appears to have gained two electrons and hence appears to have two negative charges. Therefore Oxidation number of Oxygen in H_2O is –2. Each H atom appears to have lost one electron and they appears to have one positive change. Therefore, Oxidation number of H in H_2O is +1.

In H_2 molecule, (H:H) there is equal sharing of electron pair between the two atoms. There is no net gain or loss of electron by each H atom. Therefore, Oxidation number of H in H_2 molecule is zero.

This method of counting electrons takes time therefore certain rules are framed to calculate the Oxidation number of an element. Oxidation number of an element in a compound or ion or in a molecule is computed using the following rules.

Similarly $Zn \rightarrow Zn^{2+} + 2e^-$, $Fe \rightarrow Fe^{2+} + 2e^-$, $Fe^{2+} \rightarrow Fe^{3+} + e^-$

$Sn \rightarrow Sn^{2+} + 2e$, $Sn^{2+} \rightarrow Se^{4+} + 2e$, $S^{2-} \rightarrow S + 2e$. etc.,

Reduction is defined as a reaction which involves gain of electrons by an atom, group of atoms or an ion. The substance which gains electrons is said to be reduced and is called an oxidizing agent. Reduction is reverse of Oxidation

$Zn^{2+} + 2e^- \rightarrow Zn$, $Fe^{3+} + e^- \rightarrow Fe^{2+}$, Fe^{3+} is reduction is an Oxidation agent.

$Sn^{4+} + 2e^- \rightarrow Sn^{2+}$, $S + 2e^-, \rightarrow S^{2-}$, $O_2 + 4e^- \rightarrow 2O^{2-}$, $Cl_2 + 2e^- \rightarrow 2Cl^{-1}$

Reduction and Oxidation reactions take place simultaneously. One cannot take place without other. They are referred to a Redox reactions. In these reactions transfer of electrons takes place from the reducing agent to the oxidising agent. Eg.,

$2Na_{(s)} + Cl_{2(g)} \rightarrow 2NaCl_{(s)}$

The reduction takes place as follows

$2Na \rightarrow 2Na^+ + 2e^-$ Oxidation

$Cl_2 + 2e^- \rightarrow 2Cl^-$ Reduction

$2Na + Cl_2 \rightarrow 2Na^+ Cl^-$ or 2NaCl.

In the above reaction Na loses electrons and is said to be Oxidised to Na^+ ion. The electrons lost by Na is gained by Cl_2 molecule to form Cl^- ion. Cl_2 is said to be reduced to Cl . Thus Na is a reducing agent and Cl_2 is an Oxidising agent

Similarly: $Zn \rightarrow Zn^{2+} + 2e^-$ Oxidation

$2H^+ + 2e \rightarrow H_2$ Reduction

$Zn + 2H^+ \rightarrow 2n^{2+} + H_2$

Most reactions are redox reaction (Reduction – Oxidation reactions). Few reactions are not redox reactions.

Eg: $NaOH + HCl \rightarrow NaCl + H_2O$.

Rules for calculating Oxidation number:

1. Oxidation number of element in the uncombined state or free state is Zero.

Eg: Oxidation number of H in H_2 molecule is Zero. Similarly Oxidation number of Cl in Cl_2, N in N_2, O in O_2 molecule is zero. Oxidation number of Na, Cl, Zn, Mg, Ca etc, in the uncombined state is zero. .

2. Oxidation number of F is –1 in all its compounds. Oxidation number of other halogens Cl, Br and I is –1 in all their compounds except their Oxides, Oxyacids, and Interhalogen compounds.

3. Oxidation number of Oxygen is –2 in all its compounds except OF_2 (fluorine monoxide) , super oxides and peroxides.

Oxidation number of O in OF_2 is + 2, in superoxides (O_2^-) is –2 and in peroxides (O_2^{2-}) is -1

4. Oxidation number of hydrogen is +1 in all its compounds except ionic hydrides. In metallic hydrides, Oxidation number of H is -1. Example, in NaH, LiH etc.,

5. Oxidation number of alkali metal, Li, Na, K etc, is +1 in their compounds. Oxidation number of alkaline earth metals, Ca, Ba, Mg...... is + 2 in their compounds. Oxidation number of Al is + 3 in its compounds.

6. Oxidation number of an ion is equal to charge on the ion.

Eg., Oxidation number of Al^{3+} is +3, NO_3^- ion is -1, SO_4^{-2} is -2,

NH_4^+ ion is + 1, PO_4^{3-} ion is -3 etc.,

7. The algebraic sum of the Oxidation number of all the atoms of elements in

a) a compound is Zero

b) an ion is equal to charge on ion.

Example: Calculate the Oxidation number of Mn in $KMnO_4$

Let the Oxidation number of Mn be x.

Oxidation number of K is +1 and O is – 2

$+ 1 + x - 2 \times 4 = 0$ OR $x = +7$

The Oxidation number of Mn is +7.

Calculate the Oxidation number of

i) Mn in a) K_2MnO_4 b)MnO_2 C)$MnCl_2$ d)$MnSO_4$

ii) Cr in $K_2Cr_2O_7$, $Cr_2O_7^{2-}$, $CrCl_3$, K_2CrO_4, CrO_2Cl_2, CrO_2^{2-}, CrO_4^{2-} $HCrO^{4-}$

iii) S in H_2SO_4, H_2SO_3, H_2S, SO_2, SO_3, $S_2O_3^{2-}$, $Na_2S_2O_3$, $Na_2S_4O_6$, Na_2SO_3, N_2S_2O, S_2Cl_2, $SOCl_2$,

iv) N in NH_3, N_2, N_2O, NO, NO_2, N_2O_5, HNO_2, $NaNO_3$, NCl_3,

v) P in PO_4^{3-}, $P_2O_7^{4-}$, H_3PO_4, HPO_3, HPO_2^{2-}, NaH_2PO_2, $P_2O_6^{4-}$, PH_4^+, PH_3.

vi) C in CH_4, CH_3Cl, CH_2Cl_2, $CHCl_3$, CCl_4, C_2H_6, C_2H_4, C_2H_2, diamond

$C_{12}H_{22}O_{11}$, $C_6H_{12}O_6$.

vii) Cl in ClO_3^-, ClF, ICl_3, and I in ICl_3, IF_7

viii) Tl in $TlCl_3$, H in FeO

Modern concept of oxidation and Reduction

In terms of Oxidation number, Oxidation refers to the Reduction chemical charge in which there is an increase in Oxidation number and Reduction refers to the chemical charge in which there is decrease in Oxidation number.

The reaction in which Oxidation number of an element increase is said to be oxidized and it is a reducing agent. The reaction in which Oxidation of an element decreases is said to be reduced and it is an Oxidizing agent.

Eg.: when Mg burns in air to form Mgo.

O	O		+2-2
Mg +	O_2	→	MgO

Oxidation number of Mg increases from O to + 2. Mg is Oxidized to Mg^{2+}, Mg is a reducing agent.

Oxidation number of O decreases from 0 to -2, O_2 is reduced to O^{2-}, O_2 is an Oxidising agent.

Example - 2: +2, +6, -2		O		+2, +6, -2		O
Cu SO_4	+	Zn	→	Zn SO_4	+	Cu

Oxidation number of Zn increases from O to + 2, Zn is Oxidised to Zn^{2+}.

Oxidation number of Cu decreases from +2 to 0, Cu in $CuSO_4$ is reduced or $CuSO_4$ is an Oxidising agent. Oxidation number of S and O remains the same.

Example – 3: Identify the elements and Substances undergo Oxidation and reduction and also Oxidising and Reducing agents in the following reactions.

1. $2KClO_3 \rightarrow 2KCl + 3O_2$
2. $Zn + 2HCl \rightarrow ZnCl_2 + H_2$
3. $2Na + Cl_2 \rightarrow 2NaCl$
4. $MnO_2 + HCl \rightarrow MnCl_2 + H_2O + Cl_2$
5. $H_2 + O_2 \rightarrow H_2O$
6. $C + O_2 \rightarrow CO_2$
7. $N_2 + 3H_2 \rightarrow 2NH_3$

Methods of balancing redox equations

The widely used methods for balancing chemical reactions and equations are Oxidation number change method and ion electron method.

1. Oxidation Number change method of balancing redox equations.

Following steps are involved in balancing the redox equations by this method.

1. Write down the skeleton equation

2. Write the Oxidation number of elements in each substances above the symbol of the elements.

3. Note the element which undergoes change in Oxidation number.

Note the element which shows increase in Oxidation number, ie., it is a reducing agent and the element is said to be oxidised.

Note the element which shows decrease in Oxidation number i.e., it is the oxidizing agent and is said to be reduced.

4. Note the number of units of change in Oxidation number

5. Select the suitable coefficients to the reducing agent and the Oxidising agent so that increase in Oxidation number of the elements is exactly balanced by the decrease in Oxidation number of the element reduced. Reverse the coefficients of Oxidizing and reducing agents.

6. Give appropriate coefficients to other substances in the equation so that number of atoms of any element is same on both sides of the equation.

[Balance H and O atoms when all other atoms have been balanced. Add H^+, OH^- ion or H_2O to the equations if necessary].

Examples

1. Reduction of ferric chloride by Hydrogen Sulphide

Skeleton equation is

$$\overset{+3\ -1}{FeCl_3} + \overset{+1-2}{H_2S} \rightarrow \overset{+2-1}{FeCl_2} + \overset{+1-1}{HCl} + \overset{O}{S}$$

Oxidation number of H and Cl do not change.

Oxidation number of S increases from -2 to O, this means S is oxidized.

Oxidation number of Fe decreases from +3 to +2 this means Fe^{+3} is reduced to Fe^{+2} i.e., it is a Oxidising agent.

Change in Oxidation number of S from -2 to O = -2

Change is Oxidation number of Fe from + 3 to + 2 = +1

To compensate the increase in Oxidation number of S by the decrease in the oxidation number of Fe, a coefficient of two should be given to $FeCl_3$ which means two molecules of $FeCl_3$ should be oxidized. The partly balanced equation is

$$2FeCl_3 + H_2S \rightarrow FeCl_2 + HCl + S.$$

Balanced equation is

$$2FeCl_3 \quad + \quad H_2S \quad \rightarrow \quad 2FeCl_2 \quad + \quad 2HCl \quad + \; S.$$

2. Oxidation of ammonia by Copper Oxide.

Skeleton equation is

$$\overset{+2\,-2}{CuO} \quad + \quad \overset{+3+1}{NH_3} \quad \rightarrow \quad \overset{O}{Cu} \quad + \quad \overset{O}{N_2} \quad + \quad \overset{+1-2}{H_2O}$$

Oxidation number of H and O donot change.

Oxidation number of N increases from -3 to O. NH_3 is Oxidised to N_2 i.e., this is an reducing agent.

Change in Oxidation number of Cu from + 2 to 0. This means it is an Oxidising agent.

Oxidation number decreases from +2 to 0 in Cu.

To compensate the increase in Oxidation number of N and decrease in Oxidation number of Cu, coefficients of 2 and 3 should be given to NH_3 and CuO respectively.

$$3CuO \; + \; 2NH_3 \; \rightarrow Cu \; + \; N_2 \; + \; H_2O$$

Balanced equations: $3CuO \; + \; 2\,NH_3 \; \rightarrow 3Cu \; + \; N_2 \; + \; 3H_2O$

3. Oxidation of concentrated HCl by MnO_2 Producing Chlorine

$$\overset{+4\,-2}{MnO_2} \quad + \quad \overset{+1-1}{HCl} \rightarrow \overset{+2-1}{MnCl_2} \quad + \quad \overset{+1-2}{H_2O} + \quad \overset{O}{Cl_2}$$

Oxidation number of O, H and a part of Cl in $MnCl_2$ do not change the Oxidation number. Cl increases from -1 to 0, a part of Cl or a part of HCl is Oxidised. This means HCl is a reducing agent.

Oxidation number of Mn decreases from +4 to +2, Mn or MnO_2 is reduced, MnO_2 is an Oxidation agent.

Change in oxidation number of Cl from -1 to 0 = + 1

Change in Oxidation number of Mn from +4 to +2 = -2

To Compensate the increase in Oxidation number of Cl by the decrease in Oxidation number of Mn, the number of molecules of HCl should be oxidised i.e., 1 molecule of Cl_2 should appear as the RHS of equation (a coefficient of 2 should not be given to HCl since a part of HCl is not oxidized). The equation becomes

$$MnO_2 + HCl \rightarrow MnCl_2 + H_2O + Cl_2$$

Balanced equation is: $MnO_2 + 4HCl \rightarrow MnCl_2 + 2H_2O + Cl_2$

4. Oxidation of concentrated HCl by heating with $K_2Cr_2O_7$ Crystals.

1, +6, -2   +1, -1   +1, -1   +1, -2   +1, -2   O

$$K_2Cr_2O_7 + HCl \rightarrow KCl + CrCl_3 + H_2O + Cl_2$$

Oxidation number of O, H and a part of Cl in KCl, and $CrCl_3$ do not change. Oxidation number of a part of Cl increases from -1 to O, a part of Cl and HCl is Oxidised. Therefore, HCl is reducing agent.

Oxidation number of Cr decreases from +6 to +3, Cr or $K_2Cr_2O_7$, is reduced. Therefore, $K_2Cr_2O_7$, is an Oxidising agent.

Change in Oxidation number of a part of Cl from -1 to O = +1

Change in Oxidation number of a Cr from +6 to +3 = -3 for 1 Cr atom

= -6 for 2 Cr atom

To compensate the increase in Oxidation number of Cl by the decrease in Oxidation number of Cr, six molecules of HCl should be oxidised. Or 3 molecules of Cl_2 are formed. $3Cl_2$ should appear on RHS of equation.

$$K_2Cr_2O_7 + HCl \rightarrow KCl + CrCl_3 + H_2O + 3Cl_2$$

Balanced equation is: $K_2Cr_2O_7 + 14HCl \rightarrow 2KCl + 2CrCl_3 + 7H_2O + 3Cl_2$.

QUESTIONS CARRYING ONE MARK

1) State modern periodic law.
2) What is the basis of Mendeleev's Periodic table ?

3) How many groups and periods are there in the long form Periodic table ?
4) How does electronegativity vary along a period ?
5) How many elements are present in the first transition series?
6) What is ionization energy ?
7) Give the name and symbol of the elements of the first period.
8) How many blocks of elements are there in the modern periodic table ?
9) Which of the following are representative elements? Na, Ca, Mn, Fe, Br, N, Sc.
10) Give the common electronic configurations of alkali metals.
11) What is meant by periodicity ?
12) Define electronegativity.
13) Which element has highest electronegativity ?
14) An element is found in the third period of the periodic table. Which is the outermost shell of this element ?
15) Ionisation energy of fluorine is very high. Why ?
16) Into which orbital do electrons enter in transition elements ?
17) The general electronic configuration of some elements are given below. Classify them as s - block, p - block, d - block or f - block elements:

    (a) ns^2np^5 (b) ns^1 (c) $ns^2 (n-1)d^3$

    (d) ns^2np^6 (e) $ns^2 (n-1) d^{10}np^4$ (f) $ns^2 (n-1)d^1$
18) Identify the group to which the elements with the following electronic configuration belong ?

    (a) ns^1 (b) $ns^2 np^1$ (c) $ns^2 p^4$ (d) $ns^2 np^6$
19) Which element has the most negative electron affinity value?
20) Name the scientist who stated the modern periodic law.
21) What is the basis for the modern periodic table?
22) What are the p-block elements?
23) How many series of d-block elements are present?
24) Name the element that has the highest electronegativity and the element that has the lowest electronegativity.

25) Write the atomic number of the last element in the second period and give its electronic configuration.

26) How many elements are present in the

    (i) second period (ii) fourth period ?

27) Name two

    (i) alkali metals (ii) alkaline earth metals

    (iii) transition elements (iv) noble gases

    (v) rare earth metals (vi) p – block elements

28) Mention the number of elements present in the fourth period.

29) Mention the halogen with the highest first ionization energy.

30) Mention the element has the lower ionization energy.

31) Mention which has the lower ionization energy among Mg, Mg^+.

32) Mention the type of bond formed when the bonding atoms differ largely in their electronegativity ?

33) Of the two elements bromine and chlorine, which element has the lower electronegativity ?

34) Name the smaller atom in each case:

    (a) B and Al (b) F and Na

35) Which is the smallest atom ?

36) Which element has the largest atomic radius ?

37) Which species has the higher ionic radius? Na^+ or F^-

QUESTIONS CARRYING TWO MARK

1) Draw an outline of the Periodic Table indicating the position of each block.

2) What is the basics for the classification of elements into s, p, d and f blocks ?

3) Name the block and the group to which the following elements belong. B, Mg, S and Kr.

4) Draw an outline of long form periodic table and indicate number of the groups, periods and the position of each block.

5) Define the terms: Ionisation energy, Atomic radius and Electron affinity.

6) Alkali metals has lowest first ionization energy and noble gases have highest first ionization energy values. Explain.
7) What are isoelectronic ions? Give suitable examples.
8) Which pair of atomic numbers given below has similar properties? Explain.

   (a) 3, 11 (b) 1, 11 (c) 2, 7 (d) 3, 10
9) Mg^{2+} and O^{2-} both have the same electronic configuration but still the later is larger than the former. Explain.
10) Why is ionization energy of Be is greater than that of B and O is less than that of N ?
11) Among the elements Li, Na, Mg, S and Xe which has the highest and which has the lowest first ionization energy.
12) The radius of a cation is always less than the radius of corresponding atom. Why ?
13) Which of the two, F or Cl has lower electron affinity. Why ?
14) Which species in the following groups has the smallest size ?

   (i) Na^+, K^+, Rb^+ (ii) Na, Mg, Al

   (iii) Fe, Fe^{2+}, Fe^{3+} (iv) Na^+, F^-, O^{2-}
15) Write in the increasing order of the ionic size of fluoride, nitride and oxide ions.
16) Arrange the following atoms in order of increasing radius: P, S, N. Justify your answer.
17) Arrange the following atoms in order of decreasing radius: C, Li, Be. Explain.
18) In each of the following pairs, indicate which one of two species is larger:

   (a) $N3^-$ or F^- (b) Mg^{2+} or Ca^{2+}

   (c) Fe^{2+} or Fe^{3+}. Give reason.
19) Aluminium prefer to lose 3 electrons rather than 1 and 2. Why ?
20) Which atom should have a smaller first ionization energy : oxygen or sulphur ? Give reason.
21) Which of the following atoms should have a larger first ionization energy: N or P ?

22) Why is the second ionization energy always greater than the first ionization energy for any element?
23) Alkali metals normally form ionic bond. Why ?
24) How can alkali metals and halogens attain noble gas configuration ?
25) How does electron affinity vary across a period and down a group ?
26) How does the size change when an atom is converted into: (a) an anion (b) a cation ?
27) Which among the following has highest energy? Explain (a) Na, K, Cs (b) F, Cl, I.
28) What are transition elements ?
29) Define electronegativity. How does it vary across a period and from top to bottom in a group.
30) Give a few important properties of transition elements.
31) Inert gases have zero electron affinity why ?

QUESTIONS CARRYING FOUR MARK

1) What are merits and demerits of periodic table ?
2) Write the electronic configuration of elements with atomic number 10, 11, 17 and 25 and indicate.
   (i) Which one of them is d – block element ?
   (ii) What is the nature of the bond between 11 and 17 ?
   (iii) Which one is most electropositive ?
   (iv) Which element is most stable ?
3) Define electron affinity. How does it vary in groups and periods of periodic table ?
4) Describe the periods and groups of Modern Periodic Table.
5) How does the following periodic properties of elements like atomic radius and electron affinity vary along the period and down a group in the periodic table.
6) Explain why ionization energy
   (i) Increases along a period
   (ii) Decreases along a group

7) What are covalent and ionic radii ? How does they vary in period and groups in periodic table.

8) (a) Explain S, P, d and f bock elements in periodic table.

   (b) Define Vanderwaals radius and ionic radii.

   (c) Explain the trend of electron affinity down the: (i) halogens (ii) alkali metal group.

   (d) Explain why ionization energy increases along a period.

9) (a) Describe the trends of electronegativity and electron affinity along the period and group.

   (b) Halogens cannot form ionic compounds with hydrogen. Why ?

   (c) The ionization energy of nitrogen is more than oxygen. Why ?

   (d) The second ionization energy is greater than the first why ?

4

CHEMICAL BONDING

When two or more atoms chemically combine then a bond is said to be formed between two atoms. The resulting cluster of atoms is called a molecule. The attractive force which binds the atoms in a molecule is referred to as chemical bond. A chemical bond is formed between any two atoms only if potential energy of the system is decreased. Thus the molecule has lower energy than atoms and hence molecule is stable.

The ability of the atoms to combine with each other depends mainly on the number and arrangement of electrons in their outermost shells.

Generally, it is the valence electrons involved in chemical bond formation. A bond is formed only if each atom acquires a stable electronic configuration during the process.

There are two ways in which an atom can acquire stable configuration. 1) by the transfer of electrons or 2) by sharing the electrons with other atoms. The bonding is said to be ionic if it involves electron transfer and covalent if there is sharing of electrons. In addition to these, we have Hydrogen bonding, Vander waals bonding or forces, Metallic bonding under certain specific conditions. The type of bond that may exist between any two atoms is mainly decided by the following three important properties of atoms. 1) Ionization energy 2) Electronegativity and 3) electron affinity.

Ionic Bond

An ionic bond is formed between two atoms by the transfer of one or more valence electrons from one atom to another. An ionic bond is the electrostatic force of attraction between oppositely charged ions.

Consider the formation of Lithium fluoride

The electronic configuration of Lithium is $1s^2\ 2s^1$, and that of fluorine is $1s^2\ 2s_2^2\ p^5$. When lithium and fluorine comes in contact with other, the outer $2s^1$ valence electron of lithium is transfered to the fluorine atom. We can represent this using the Lewis dot symbols.

$$\cdot Li \quad + \quad :\ddot{F}: \longrightarrow Li^{+}\ :\ddot{F}: \qquad \text{or LiF}$$

$$1s^2\ 2s^1 \qquad 1s^2\ 2s^2 2p^5 \qquad 1s^2,\ 1s^2 2s^2\ 2p^6$$

The $Li^+\ F^-$ unit is held together by the electrostatic force of attraction between the positively charged lithium and the negatively charged fluoride ion. The result of this attraction is ionic bond.

Note: Consider the formation of NaCl

$$Na \ + Cl \longrightarrow Na^+ + Cl^-$$

$$2,8,1 \quad 2,8,7 \qquad\qquad 2,8 \quad 2,8,8.$$

Lattice energy

Lattice energy (U) of an ionic crystal is the energy evolved when one mole of the crystal is formed from gaseous ions

$$Na^+(g) + Cl^-(g) \rightarrow NaCl(s) \qquad \Delta H_U = -780\,kJmol^{-1}$$

Lattice energy cannot be measured directly, but the experimental values can be obtained from thermo chemical data using born Haber cycle. Theoretical values for lattice energy may be calculated using Born - Lande equation.

The born Haber cycle used in determination of Lattice energy is represented as follows.

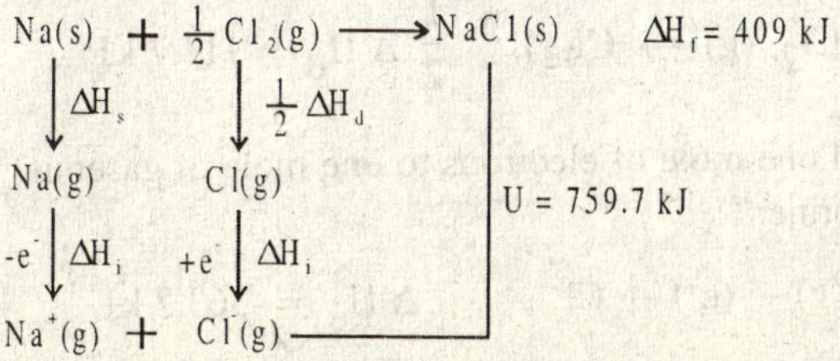

Fig 4.1 : Born – Haber Cycle

Thus the lattice energy can be calculated according to Hess's law, which states that "the enthalpy of formation of NaCl is same whether the reaction is taking place in one step or several steps" hence, we can write.

$$\Delta H_f = \Delta H_s + \Delta H_i + \frac{1}{2}\Delta H_d + \Delta H_e + U$$

$$\therefore U = \Delta H_f - \Delta H_s - \Delta H_i - \frac{1}{2}\Delta H_d - \Delta H_e$$

Where ΔH_f = Enthalpy of formation

ΔH_s = Enthalpy of sublimation

ΔH_i = Enthalpy of ionization

ΔH_d = Enthalpy of dissociation

ΔH_e = Enthalpy of electron affinity

U = Lattice energy.

The Born Haber cycle can be explained as follows

I Step — Conversion of solid sodium to sodium vapour

$Na(s) \rightarrow Na(g)$

$\Delta H_s = 108.5$ kJ.

II Step — Ionisation of one mole of gaseous sodium atoms

$Na(g) \rightarrow Na^+ + e^-$ $\quad \Delta H_i = 492.7$ kJ.

III Step — Dissociation of $\frac{1}{2}$ mole of gaseous chlorine to atomic chlorine

$\frac{1}{2} Cl_2(g) \rightarrow Cl(g), \quad \frac{1}{2}\Delta H_d = 112.7$ kJ.

IV Step — Add one mole of electrons to one mole of gaseous chlorine

$Cl(g) + (e^-) \rightarrow Cl^-$ $\quad \Delta H_e = -363.2$ kJ.

V Step Combine one mole of gaseous Sodium and one mole of gaseous chlorine to form one mole of NaCl solid

$Na^+ (g) + Cl^- (g) \rightarrow NaCl$ $\Delta H_U = ?$

Reverse the step 5 that give the lattice energy

$NaCl(s) \rightarrow Na^+ (g) + Cl^- (g)$

Thus ΔH_U of NaCl is -759.7 kJ

Condition for the formation of ionic bond

1) Low ionisation energy: Ionization energy should be low such that only small amount of energy is required to release its valence electrons. Then it readily forms cation (x+)

2) High electron affinity: The electron affinity of the non metal should be high such that the atom may accept the electron more readily to form anion (x^-)

3) High lattice energy: Strong electrostatic force of attraction is involved between the cation and the anion. In an ionic crystal Lattice energy is a measure of its stability. Greater the force of attraction, greater is the amount of energy released.

4) Large difference of electro negativity: Complete transfer of electrons takes place only when the atoms in an ionic crystal possess large difference in electro negativity of anion and cation.

The Covalent bond

G.N. Lewis proposed the theory of covalent bond formation in the molecules like H_2, Cl_2, O_2 etc. A covalent bond is formed by the sharing of a pair of electrons between the two atoms, each atom contributing one electron to the shared pair. By sharing this bonded pair, each of two combining atoms attain stable electronic configuration. A covalent bond is formed when the combining atoms have comparable electro negativities. For the formation of a covalent bond, the atom of the element should have one or more unpaired electrons. Number of covalent bonds formed is equal to number of unpaired electrons. A covalent bond formed by the sharing of one electrons pair is called a single bond represented by writing a single solid line. A covalent bond formed by the sharing of two pairs of electrons is a double bond and three pairs of electrons is a triple bond.

When a covalent bond is formed, the two electrons of the shared pair have opposite spins and are attracted by the nuclei of both the atoms. The attractive forces bind the two atoms together. Unshared pairs of valence electrons are called lone pairs. The Lewis octet theory was the first explanation of a covalent bond represented in the form of electron dot formula.

Let us consider the formation of hydrogen molecule (H_2), which involves bonding between two H-atoms. The single valence electron of two hydrogen atoms is shared for the formation of a covalent bond. As a result of this sharing, both the atoms attain stable electronic configuration of helium.

$$H^{\bullet} + H^{\bullet} \rightarrow \underset{\text{Shared Pair}}{\underset{\uparrow}{H:H}} \quad \text{or} \quad \underset{\text{Covalent bond.}}{\underset{\uparrow}{H-H}}$$

A covalent bond is indicated by a dash(-) between the two atoms. In fact, the positive nuclei of two atom are pulled towards each other by the attraction of shared electron pair. At the same time, the nuclei of two atoms also repel with each other as the electron does. It is the net attractive force between the shared electrons and the nuclei that holds the atoms together. Therefore an alternative definition for a covalent bond is "The attractive force between atoms combined by sharing of an electron pair". The compounds formed by this way are called covalent compounds.

Conditions for the formation of covalent bonds

The conditions favourable for the formation of covalent bond are

1. **The number of valence electrons.**

Each atom should possess 5,6, and 7 valence electrons so that both achieve the stable octet by sharing 3,2, or 1 electron pair

2. Equal electronegetivity

The atoms with equal electronegetivity will not have transfer of electrons hence they go for pairing and sharing of electrons

3. Equal sharing of electrons

The two atoms combining should have equal electron affinity so that they attract the bonding electron pair equally. Thus equal sharing of electrons will form a non polar covalent bond.

Pictorial representation of Covalent bond

HCl, NH_3, CO_2 and N_2.

1. H :C̈l: or H – Cl Hydrogen chloride

2. 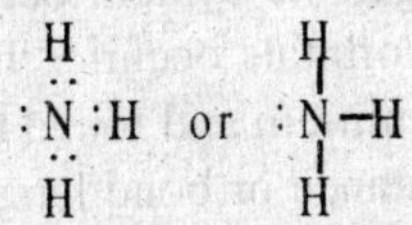

Ammonia

3. O = C = O or :Ö : C :Ö :

Carbon dioxide

4. N ≡ N or N ⋮ N Nitrogen

Valence bond theory

To explain the formation of covalent bond, Pauling and Slater (1931) used the valence bond theory. According to valence bond theory, a covalent bond is formed by the overlap of atomic orbitals which contain unpaired electrons. The resulting bond consists of pair of electrons with opposite spins, which occupy the atomic orbitals of bonded atoms. As a result of overlap of atomic orbitals, a molecular orbital is formed. The molecular orbital is localized between the two atom which are involved in bonding. The molecular orbital contains pair of electrons. The maximum density of the electrons are found between the two atoms. The formation of a covalent bond leads to decrease in the energy of the system. The atoms involved in bonding loses energy and reaches a minimum value when they approach the bonding distance. Due to this energy, the system is stabilised.

Formation of hydrogen molecule can be explained according to valence bond theory as follows. When the hydrogen atoms are in isolated state, there is no force of attraction between them. The potential energy considered to be zero. As the atoms approach each other, their potential energy decreases due to the influence of electron cloud of one atom on the nucleus of the other atom. At a certain minimum energy, the force of attraction between the nuclei and electrons of two atoms is exactly balanced by the repulsive forces between the two atoms. At this stage, the system becomes most stable and maximum overlapping of orbitals occurs. The distance between the two hydrogen atom is minimum and found to be 75 pm ($0.75A^0$). This is known as bond distance or bond length.

HYBRIDISATION

Hybridisation involves the linear combination of two or more atomic orbitals of different shape and energy to form an equal number of new identical orbitals of same shape and energy. Atomic orbitals involved are of comparable energies lying in the same principal energy level but occasionally is adjacent energy levels. The new equivalent orbitals formed are called hybrid orbitals. The hybrid orbitals formed have one lobe much larger than the other lobe, therefore, the electron density is concentrated on one side of the nucleus and the bond formed with such an orbital is stronger than the one formed with either pure s or pure p orbitals.

The hybrid orbitals are one directional in character. They arrange symmetrically in such a way that they are as far apart as possible so that there is minimum repulsion between them. The directional character of the hybrid orbitals is responsible for definite shape or geometry of the molecule.

Chemical bond is formed by the overlapping of the hybrid orbitals.

The different types of hybridisation is possible by the mixing of s, p and d orbitals and the shape of the resulting molecule of all the hybrid orbitals have shown as follows.

Type of hybridization	Shape of resulting molecule
1. sp hybridisation	Linear
2. sp^2 hybridisation	Trigonal planar
3. sp^3 hybridisation ($s + p_x + p_y + p_z$)	Tetrahedral
4. dsp^2 hybridisation ($s + p_x + p_y + dx^2 - y^2$)	Square planar
5. dsp^3 or sp^3d hybridisation ($s + p_x + p_y + p_z + d_z{}^2$)	Trigonal Bipyramidal
6. dsp^3 or sp^3d hybridisation ($s + p_x + p_y + p_z + dx^2 - y^2$)	Square Pyramidal
7. d^2sp^3 hybridisation ($S + p_x + p_y + p_z$ + or $sp^3 d^2$) hybridisation $dx^2 - y^2 + d_2{}^2$	Octahedral

If the hybrid orbitals have lone pairs, symmetry of the molecule will be lowered.

1. sp hybridization : sp hybridization involves the combination of one s and p orbitals. This results in the formation of two hybrid orbitals which lie at an angle of 180^0 to each other. The sp hybrid orbital has 50% s character and 50% P character. The bonds formed by two hybrid orbitals are at 180^0 to each other and the resulting molecule is linear.

Eg., $BeCl_2$ molecule

Be: $z = 4, 1s^2 2s^2$ — no unpaired electrons in the ground state

$1s^2 2s^1 2p^1$ — two unpaired electrons in the exited state

Two sp hybrid orbitals with two bonded Pairs of electrons.

Cl $z = 17$ - - - - - - - - $3s^2\ 3p^5$, one unpaired electron in one of the p orbitals.

If in Be atom s and p orbitals are not hybridized, then bonds formed with the two Cl atoms will have different bond length, since one bond will be formed by s – p overlapping and the other by p – p overlapping. In $BeCl_2$ molecule the two bonds have the same strength and length. In order to account for the similarity of the two Be – Cl bonds, the concept of hybridization is used.

Sp hybridation in Be atom results in the formation of two hybrid orbitals which are at an angle of 180^0 to each other. Each of the half filled hybrid orbitals of Be atom overlap with half filled P orbital of one chlorine atom along the nuclear axis to form two Be –Cl σ bonds. $BeCl_2$ is a linear molecule. Cl Be Cl bond angle is 180^0.

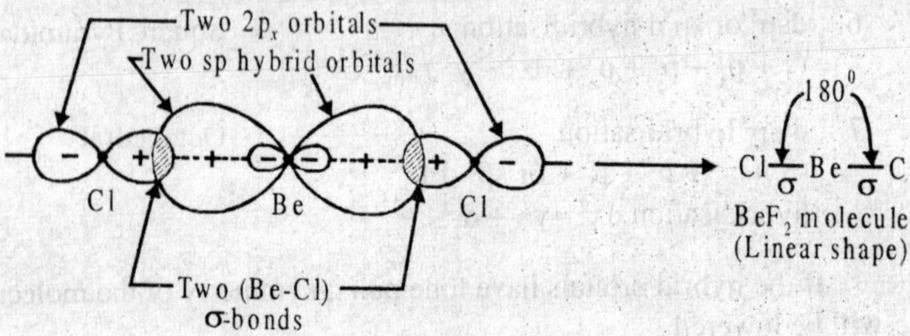

Fig 4.2: Structure of Beryllium Chloride

2. sp^2 hybridisation: This type of hybridization involves the combination of one s and two p orbitals. This results in the formation of three hybridized orbitals which are at an angle of 120^0 to each other and are directed towards the corners of a triangular plane. The sp^2 orbital has 33% s character. The bonds formed by these hybrid orbitals are also at 120^0 and hence the resultant molecule has a trigonal planar shape. Eg., BCl_3 molecule (or BF_3)

B, $z = 5$ $1s^2\ 2s^2\ 2p^1$ one unpaired electron in ground state

$1s^2\ 2s^1\ 2p^2$ three unpaired electrons in excited state.

With sp^2 hybridisation of B atom, three hybrid orbitals which are 120^0 to each other are formed. Each of the half filled hybrid orbital of the atom overlaps with half filled P orbital of one chlorine atom along the nuclear axis to form three B – Cl σ bonds. Thus BCl_3 is a triangular planar molecule. Cl B Cl bond angle is 120^0. B atom is at the centre and the three Cl atom are at the corners of triangle.

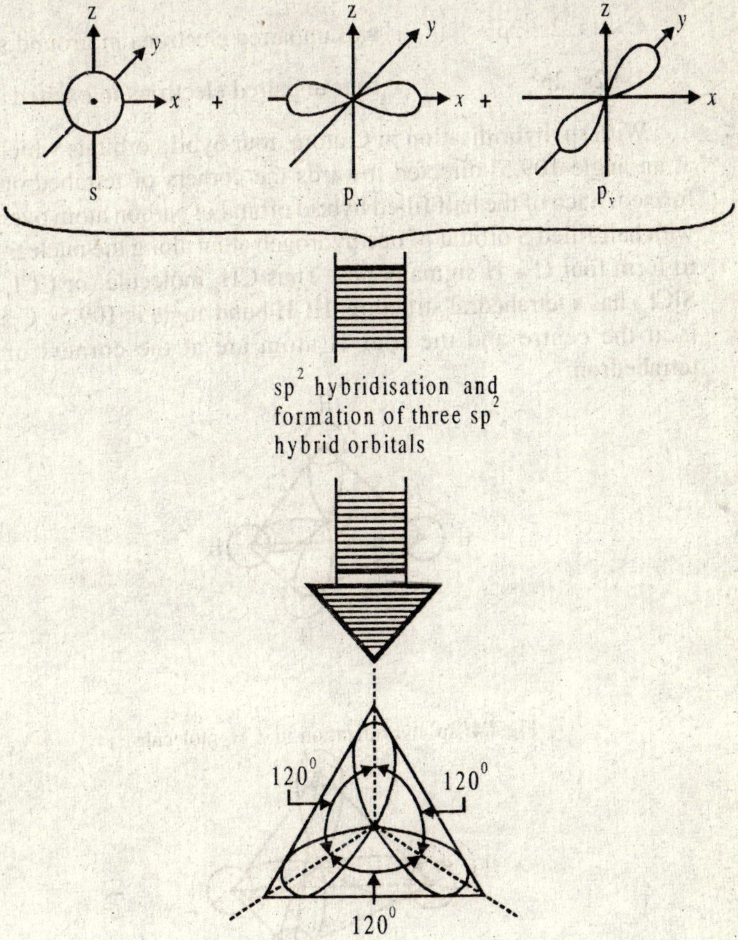

Fig 4.3: Formation of sp² Hybrid Orbitals

3. sp³ hybridisation: Involves the combination of one s and three p orbitals, resulting in the formation of four hybrid orbitals which are directed towards the corners of a tetrahedron and are at an angle of $109^0\ 28^1$ (or approximately 109.50^0). This angle is tetrahedral angle. The sp^3 hybrid orbital has 25% s character. The bonds formed by the hybrid orbitals are at an angle of 109.5^0, and the resultant molecule has a tetrahedral shape.

Eg., CH_4, CCl_4 or $SiCl_4$.

C - $1s^2\ 2s^2\ 2p^2$	two unpaired electrons in ground state
$1s^2\ 2s^1\ 2p^3$	four unpaired electrons in excited state

With sp^3 hybridisation in C atom, four hybrid orbitals which are at an angle 109.5^0 directed towards the corners of tetrahedron are formed. Each of the half filled hybrid orbital of carbon atom overlaps with half filled S orbital of one hydrogen atom along the nuclear axis to form four C – H sigma bonds. Thus CH_4 molecule (or CCl_4 and $SiCl_4$) has a tetrahedral structure. HCH bond angle is 109.5^0. C atom is at the centre and the four H atom are at the corners of the tetrahedron.

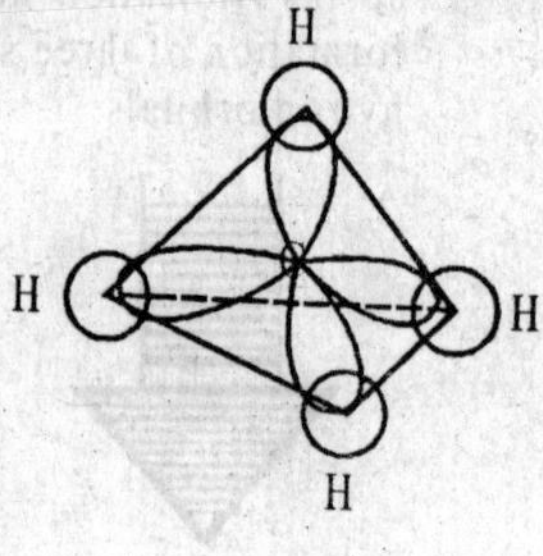

Fig 4.4: sp^3 hybridization in CH_4 molecule

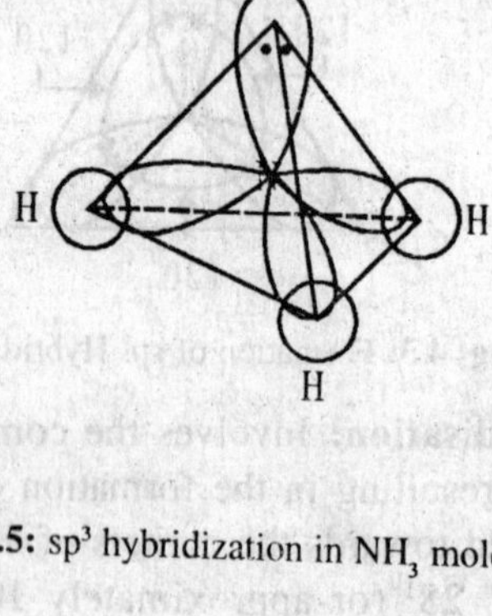

Fig 4.5: sp^3 hybridization in NH_3 molecule

1. sp^3d hybridisation or d sp^3 hybridisation : This type of hybridization Involves the combination of one s, three P and one d orbital ($d_z{}^2$ orbital). This result in the formation of five hybrid orbitals which are directed towards the corners of a trigonal bipyramid. These hybrid orbitals are not equivalent. Three of the hybrid orbitals are directed towards the corners of a triangular plane and are at an angle of 120^0 to each other, and are considered as sp^2 hybrids. They are called equatorial hybrid orbitals and the bonds formed by them are equatorial bonds. The other two hybrid orbitals are 180^0 to each other i.e., they lie one above one below the triangular plane at an angle of 90^0 to the plane. They are considered as dp hybrids. They are called axial orbitals and the bonds formed by them are axial bonds. Equatorial bonds are stronger and shorter than the axial bonds.

Eg., PCl_5

P, z = 15	- $1s^2\ 2s^2\ 2p^6$	$3s^2\ 3p^3\ 3d^0$	three unpaired electrons in ground state
		$3s^1\ 3p^3\ 3d^1$	five unpaired electrons in excited state.

With sp^3d hybridisation in P atom, five hybrid orbitals which are directed towards the corners of a trigonal bipyramid are formed. Three of the hybrid orbitals are at an angle of 120^0 to each other and are directed towards the corners of a triangular plane (equatorial orbitals) one hybrid orbital is above and one hybrid orbital is below the plane at an angle of 90^0 to the plane. (axial orbitals)

Each of the half filled hybrid orbital of phosphorous atom overlap with half filled p orbital of one chlorine atom along the nuclear axis to form five P – Cl σ bonds. The P – Cl bonds are not equivalent. P – Cl bonds in the triangular plane are equatorial bonds and are at 120^0. The P – Cl bonds perpendicular to the plane are axial bonds and are at 90^0 to the plane

PCl_5 molecule has trigonal bipyramidal structure with P atom at the centre and the five Cl atoms at the corners of the trigonal bipyramid.

PCl_5 molecule has unsymmetrical structure since some bond angles are 120^0 and same are 90^0. Such unsymmetrical structures are unstable and hence PCl_5 molecule is reactive.

2. sp^3d^2 hybridisation or d^2sp^3 hybridisation: Involves the combination of one s, three P and two d orbitals ($dx^2 - y^2$ and dz^2). This results in the formation of six hybrid orbitals which are directed towards the corners of an Octahedron. Four of the hybrid orbitals are directed towards the corners of a square plane and are at an angle of 90^0 to each other. The other two hybrid orbitals lie one above and one below the square plane and are at an angle of 90^0 to the plane. All the six hybrid orbitals are equivalent.

Eg., SF_6 molecule (Sulphur hexafluoride)

S, z = 16. $1s^2\ 2s^2\ 2p^6$	$3s^2\ 3p^4\ 3d^0$	two unpaired electrons in ground state
	$3s^1\ 3p^3\ 3d^2$	six unpaired electrons in excited state
F, z = 9	$1s^2\ 2s^2\ 2p^5$	one unpaired electron is in the P orbital.

With sp^3d^2 hybridisation in sulphur, six hybrid orbitals which are directed towards the corners of an octahedron are formed. Each of the half filled hybrid orbital of s atom overlap with half filled hybrid P orbital of one F atom along the nuclear axis to form six S – F sigma bonds. SF_6 molecule has an octahedral structure with sulphur at the centre and six F atom at the corners of the Octahedron. FSF bond angles equal to 90^0.

3. sp^3d^3 hybridisation: Involves the combination of one s, three p and three d orbitals. This results in the formation of seven hybrid orbitals which are directed towards the corners and a Pentagonal bipyramid. Five of them are directed towards the corners of a pentagonal plane (equatorial hybrid orbitals) and two of them lie one above and one below the plane (axial hybrid orbitals). At 90^0 to the plane.

Eg., IF_7 molecule (Iodine heptafluoride).

I : z = 53 - - - - - - -- $5s^2\ 5p^5$ one unpaired electrons is ground state

$5s^1\ 5p^3\ 4d^3$ seven unpaired electrons is excited state

Each of the half filled hybrid orbital of Iodine atom overlaps with half filled p orbital of F atom lie on the nuclear axis to form seven I – F bonds. IF_7 molecule has pentagonal bipyramidal structure.

In all the above examples, the hybrid orbitals of central atom have only bonded pairs of electrons.

Examples of molecules with the central atom having, bonded pairs as well as lone pairs in hybrid orbitals.

a) NH_3 molecule has a trigonal pyramidal structure with a lone pair of electrons on nitrogen atom. N atom is at the apex and the three H atoms are at the corners of the triangular base of the pyramid.

With sp^3 hybridisation in N atom four hybrid orbitals which are directed towards the corners of a tetrahedron are formed. Three of them are half filled and one of them is completely filled i.e., has a lone pair electrons. Each of the half filled hybrid orbital of N atom overlap with half – filled s orbital of one hydrogen atom along the nuclear axis to form three – N – H σ bonds.

b) H_2O molecule

With sp^3 hybridisation in Oxygen atom four hybrid orbitals which are directed towards the corners of a tetrahedron are formed. Two of them are half filled and the other two of them are completely filled. Each of the half filled hybrid orbital overlaps with half filled s orbital of hydrogen atom along the nuclear axis to form two O – H σ bonds.

H_2O molecule has bent or angular structure with two lone pairs of electrons on oxygen atom.

HNH bond angle in NH_3 molecule is found to be 107.3^0. The deviation from tetrahedral angle of 109.5^0 is due to the presence of a lone pair of electrons on nitrogen atom.

HOH bond angle in H_2O molecule is bond to be 104.5^0. The greater deviation from the tetrahedral angle is due to the presence of two lone pairs of electrons or oxygen atom.

(c) PCl_3 molecule

P - - —— - - - - - $3s^2 3p^3$. Sp^3 hybridisation

The structure of PCl_3 molecule is similar to that of NH_3 molecule.

Pure atomic orbitals of the central atom	Hybridization of the central atom	Number of hybrid orbitals	Shape of hybrid orbitals	Examples
s, p	sp	2	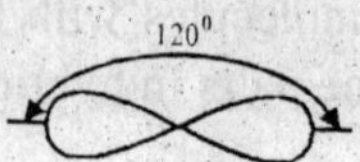 Linear	$BeCl_2$
s, p, p	sp^2	3	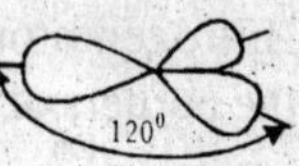 Planar	BF_3
s, p, p, p	Sp^3	4	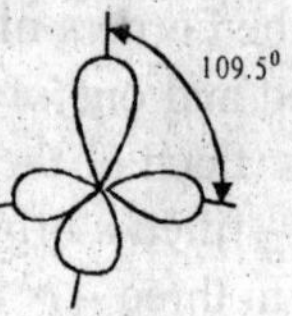 Tetrahedral	CH_4, NH_4,
s, p, p, p, d	Sp^3d	5	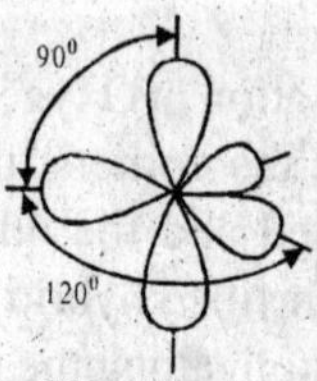 Trigonal bipyramidal	PCl_5
s, p, p, p, d, d	sp^3d^2	6	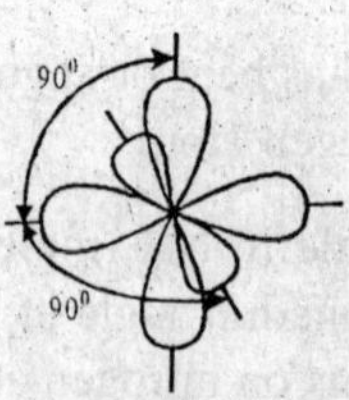 Octahedral	SF_6

Fig 4.6 Type of Hybrid orbitals and their respective shapes

Multiple bonds

Multiple bonds involve the formation of σ as well as π bonds [formed by lateral overlapping of p or d orbitals]

With sp^3 hybridisation in each C atom. Four hybrid orbitals which are directed towards the corners of a tetrahedran are formed. Each of them is half filled. One half filled hybrid orbital of each C atom overlap naturally along the nuclear axis to form C – C σ bond. Each of the remaining three half filled hybrid orbitals of each C atom overlap with half filled S orbital of one H atom along the nuclear axis to form Six C – H σ bonds. C_2H_6 molecule is not a planar molecule. There is tetrahedral distribution of atoms around each c atom. HCH and HCC bond angle is 109.5^0.

C –C σ bond is formed by the overlapping of sp^3 hybrid orbitals. C- C bond length is 154 pm.

Eg., Ethylene molecule (C_2H_4)

Each C atom is sp^2 hybridised with one of the P orbitals remaining unhybridised (say P_z orbital)

With sp^2 hybridisation in each C atom Three hybrid orbitals which are directed towards the corners of an equilateral triangle and which are at an angle of 120^0 to each other formed. One half filled hybrid orbital of each C atom overlap along the nuclear axis to form a C – C σ bond. Each of the other two half filled hybrid orbitals and each C atom overlap with half filled s orbital of one H atom along the nuclear axis to form four C – H σ bond. The unhybridised p orbital (p_z orbital) of the two c atoms overlap laterally above and below the nuclear axis to form a π bond. C_2H_4 molecule is a planar molecule. CCH and HCH bond angle is approximately 120^0. The two C atoms are held together by a double bond consisting of one σ bond and one π bond. C – C σ bond in ethylene is by the overlapping of sp^2 hybrid orbitals. C – C bond length is 135 pm.

Acetylene molecule – C_2H_2

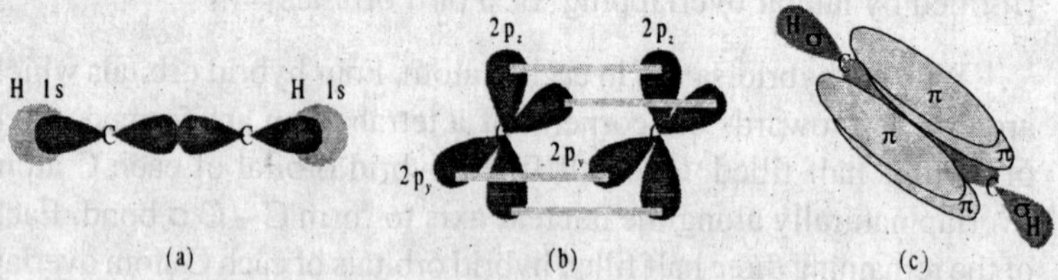

Fig 4.7 : Hybridization in Acetylene molecule

Each C atom is sp hybridized with two of the P orbitals remaining unhybridised (say P_y and P_z orbitals).

With sp hybridization in each C atom, two hybrid orbitals are at an angle of 180^0 to each other are formed. One of the half filled hybrid orbital of each C atom overlap mutually along the nuclear axis to form a C – C σ bond. The other half filled hybrid orbital of each C atom overlap with half filled S orbital of one H atom along the nuclear axis to form two C – H σ bonds. Unhybridised P_z orbital of the two c atom overlap laterally to form a π bond. Unhybridised P_y orbital of the two C atom also overlap laterally to form another π bond. C_2H_2 is a linear molecule. HCC bond angle is 180^0. The two C atoms are held together by a triple bond consisting of a σ bond and two π bonds.

C – C σ bond is formed, by the overlapping of sp hybrid orbitals. C – C bond length is 121 pm.

With increasing s character, C – C bond strengths and bond energies increase in the order

$C_2H_6 < C_2H_4 < C_2H_2$.

Resonance :

Resonance forms of Hydrogen molecule and Benzene.

A Resonance is one of two or more Lewis structures for a single molecule that cannot be described by only one Lewis structure. The symbol "⟷" indicates that the structures shown are resonance structures. The term resonance means that use of two or more Lewis structures to represent a particular molecule.

1. H_2 Molecule :

According to Heither – London theory of H_2 molecule, it was found that H_2 molecule can be represented by the pure covalent structure and ionic structure. However, it is true that H_2 molecule is a resonance hybrid of the following three structure.

$$H_A - H_B \longleftrightarrow H_A^- - H_B^+ \longleftrightarrow H_B^+ - H_A^-$$

2. Benzene :

The concept of resonance applies equally to organic systems. A well known example is Benzene molecule.

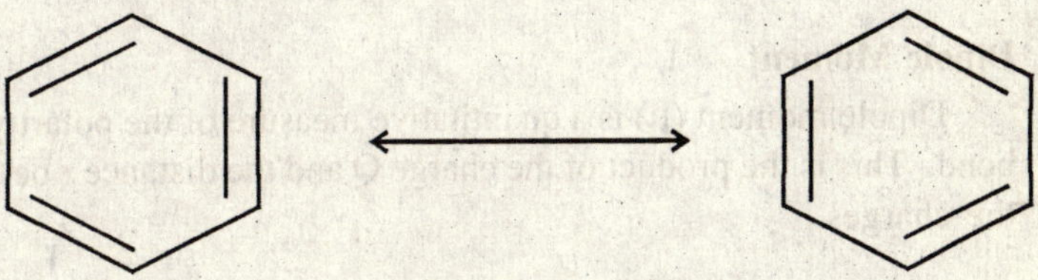

The distance between the adjacent 'C' atoms in benzene is 140 Pm, which is between the length of a C – C bond and C = C bond i.e., 133 Pm). Resonance forms have almost the some energy. Thus a resonance form having higher energy than the actual resonance forms cannot be taken as contributing structure.

Polar molecules :

A polar molecule is made by polar covalent bond. A covalent bond in which electrons are shared unequally and the bonded atoms acquire a partial positive and negative charge is a polar covalent bond.

H – Cl or H : Cl

Due to greater attraction of one nucleus (Cl) for the electrons, the shared pair is displaced toward it. This makes one end of the bond partially positive and the other end partially negative.

Some more example of polar molecules

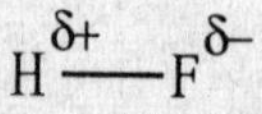

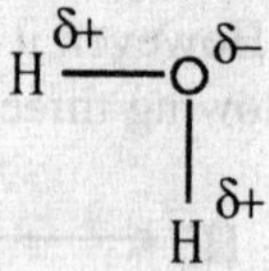

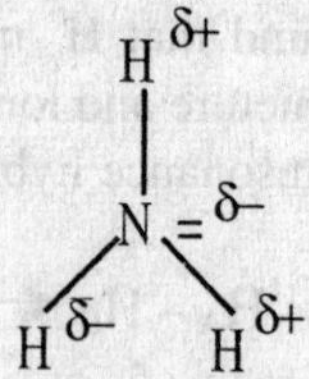

Hydrogen fluoride **Water** **Ammonia**

Greater the deference in electronegetivity between two atoms, greater the polarity. The amount of polarity of a bond is determined by the difference of electronegativity of the bonded atoms.

Dipole Moment

Dipole moment (μ) is a quantitative measure of the polarity of a bond. This is the product of the charge Q and the distance r between the charges.

$$\mu = Q \times r$$

To maintain electrical neutrality, the charges on both the ends of an electrically neutral diatomic molecule must be equal in magnitude and opposite in sign. However the quantity Q refers to only magnitude and not its sign. Therefore, μ is always positive.

Dipole moments are usually expressed in debye units (D)]

$$1D = 3.33 \times 10^{-30}\ Cm \quad \text{where } C = \text{coulomb, m - meter}$$

Diatomic molecules containing atoms of the same element (Ex.: H_2, O_2, Cl_2 etc) do not have dipole moment, where as diatomic molecules containing atoms of different elements (Ex.: HCl, CO and NO) have dipole moments.

The dipole moment is a vector quantity. Which means it has both magnitude and direction. The measured dipole moment is vectorial sum of the bond moments.

Application of dipole moment

1. Ammonia molecule

Ammonia molecule has pyramidal structure. The dipole moment of three N – H bonds and the contribution of lone pair on nitrogen atom to dipole moment give the resultant dipole moment of NH_3 molecule as 1.49D or 4.9 x 10^{-30} cm.

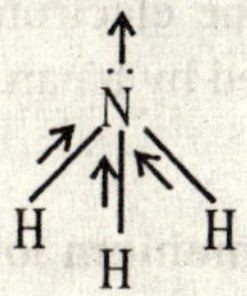

2. Water molecule (H_2O)

Water molecule has H – O – H bond angle 104.5^0. It has two lone pair of electrons on oxygen atom. The dipole moment of water is the resultant of two O – H bonds inclined at 104.5^0, and two lone pairs of electrons. The net dipole moment is found to be 1.84 D or 6.1 x 10^{-30} Cm.

3. Carbon dioxide molecule

Carbon dioxide is a linear molecule. The arrows show the shift of electron density from the less electronegative carbon atom to more electronegative oxygen atom. The dipole moment of the entire molecule is made up of two bond moments. The two bonds moments in CO_2 are equal in magnitude, since they point in opposite directions in a linear molecule. Therefore, the net dipole moment for CO_2 is zero.

Co - ordinate bond:

A Covalent bond results from the sharing of a pair of electrons between two atoms, where each atom contributes one electron to the bond. It is also possible to have an electron pair bond where both the electrons originate from one atom and none from the other atom. Such bonds are called co-ordinate bonds or co-ordinate covalent bonds or dative bonds. Since, in co-ordinate bonds, two electrons are shared by two atoms. They differ from normal covalent bonds in the way they are formed and hence formed bonds are identical to normal covalent bonds.

A co ordinate bond is established between two atoms, when one of which has a complete octet and at the same time possess at least one lone pair of electrons while the other is short of two electrons. The former atom contributes one such pair of electrons for mutual sharing between the two atoms as a result of which second atom also completes its octet and acquires a stable noble gas configuration. The atom which contributes the pair of electrons in the donor and the atom which accepts the electron pair is the acceptor. A co-ordinate bond is represented by an arrow mark pointing towards the acceptor.

Eg: **formation of ammonium ion: [NH_4^+]**

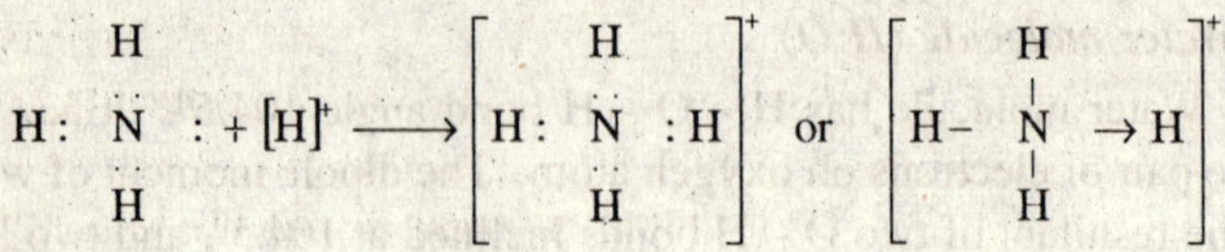

NH_3 has three covalent bonds and a lone pair. It reacts with H^+ to form NH_4^+ . The fourth bond is equivalent to the other three bonds, but it is formed by sharing an electron pair from N with H^+. This bond is the coordinate bond. It cannot be distinguished from the other three bonds, all four bonds are equivalent for NH_4^+.

2. Formation of an adduct

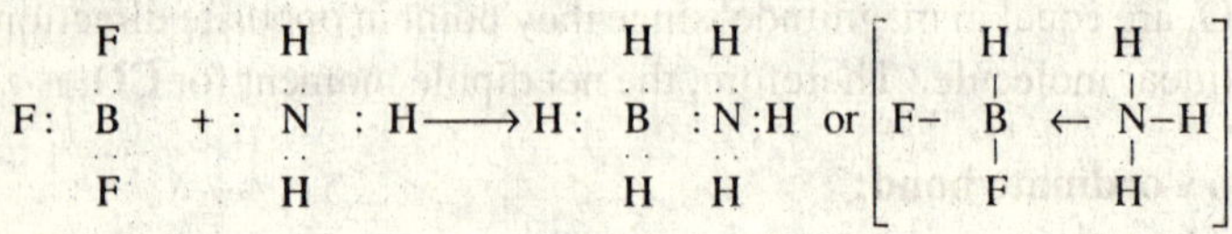

Nitrogen of NH_3 donates the lone pair to electron deficient B-atom in BF_3 when coordinate bond is formed between the two atom. By the sharing of their electron pair, B-atom completes its octet of electrons. Formal charges on bonded N atoms is-1 and +1 respectively.

3. $H_2O + H^+ \longrightarrow H_3^+O$ Hydronium ion.

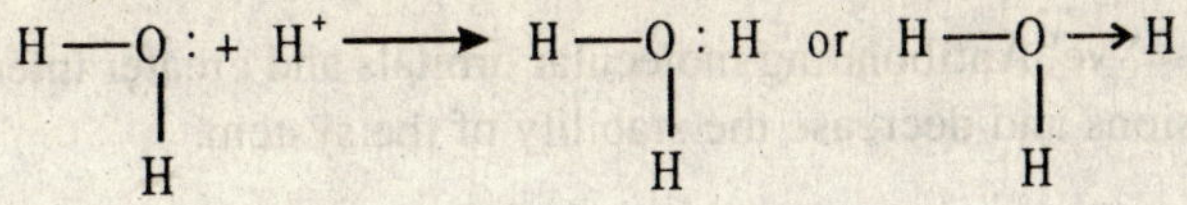

Molecular orbital theory

This theory was proposed by Hund mullikan and Huckel to explain the covalent bond.

It depends on the atomic orbitals of combining atoms, these atomic orbitals combine with each other to give molecular orbitals. The atomic orbitals give rise to bonding, antibonding and non -bonding molecular orbitals. Consider two atoms A and B each possessing a single valence electron. The valence electrons will be in region where the orbital over lap and can associate with either nucleus. Each atomic orbitals and also molecular orbitals are associated with a wave function ψ.

Let ψ_A and ψ_B be the wave function associate with the atomic orbitals of A and B. They combine together to give molecular orbitals on which one is bonding(ψ_a) and other is antibonding (ψ_b)

$$\psi = \psi_A + \psi_B$$

$$\psi_a = \psi_A - \psi_B$$

The electrons are moving in a molecular orbitals whose total volume almost equal to sum of the volume of atomic orbitals of A and B.

$$\psi_b^2 = \psi_A^2 + \psi_B^2 + 2\psi_A\psi_B$$

$$\psi_a^2 = \psi_A^2 + \psi_B^2 + 2\psi_A\psi_B$$

The integral $S = \int_0^{\infty} \psi_A\psi_B \, dT$ is called as overlap integral and it gives the electron density in space. S= + ve, bonding molecular orbitals and have less inter nuclear repulsion due to shielding of nuclei by increased electron density and stable.

S = -ve, Antibonding molecular orbitals and greater internuclear repulsions and decrease the stability of the system.

S = O, non - bonding molecular orbital and it is intermediate combination between bonding and antibonding.

$$\int_0^\infty \psi_b^2 d\psi = \int_0^\infty \psi_a^2 d\psi = \int_0^\infty \psi_A^2 d\psi = \int_0^\infty \psi_B^2 dT = 1.$$

Therefore all are normalised functions. The probability of finding electron in space is unity. Multiply ψ_a andψ_b by corresponding normalising factorN_b and N_a

$\psi_b = \psi_A + \psi_B$

$\psi_a = \psi_A - \psi_B$

$$\int_0^\infty \psi_b^2 d\psi = N_b^2 \int_0^\infty \psi_n^2 d\psi + N_b^2 \int_0^\infty \psi_b^2 d\psi = 2N_b^2 \int_0^\infty \psi_b^2 \psi_A \psi_B \, d\psi$$

$1 = N_b^2 \; [1 + 1 + 2S]$

$1 = N_b^2 \; [2 + 2S]$

$$N_b = \frac{1}{\sqrt{2+2S}} = \sqrt{\frac{1}{2(1+S)}}$$

Similarly $N_a = \sqrt{\frac{1}{2(1+S)}}$

Since 'S' is very small, $N_a = N_b = \frac{1}{\sqrt{2}}$

$$\therefore \psi_b^2 = \frac{1}{\sqrt{2}}(\psi_A + \psi_B)$$

$$\psi_a^2 = \frac{1}{\sqrt{2}}(\psi_A - \psi_B)$$

Consider two atoms of A.

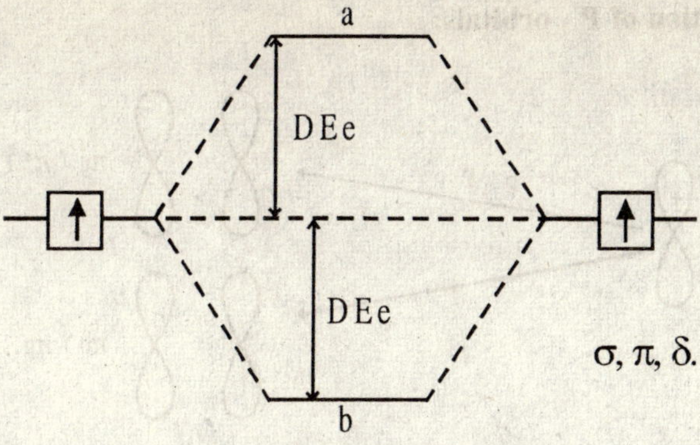

Fig 4.8 : Molecular orbital diagram

ΔE_c is referred as the exchange energy i.e., it is the energy difference between bonding molecular orbital or antibonding molecular orbital and the atomic orbitals of combining atoms. This is also referred as "bond energy". If ψ of the combining atoms is same then ΔE_c is same for bonding molecular orbital and antibonding molecular orbital

Example Homo nuclear diatomic molecules.

If ψ of the combining atoms are different then, ΔE_c is different for bonding molecular orbital and antibonding molecular orbitals.

Example Homo nuclear diatomic molecules.

Combination of s - orbitals:

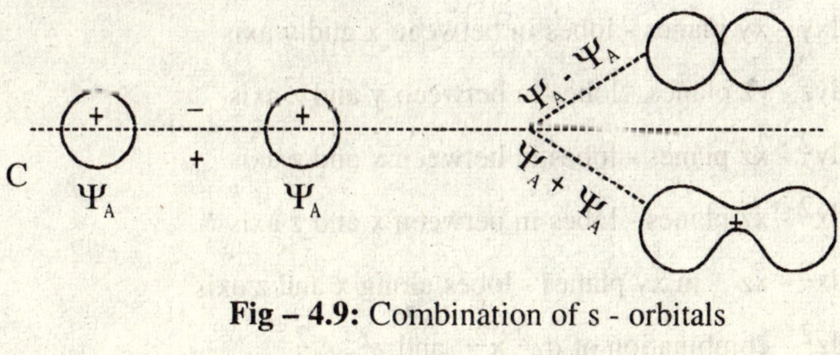

Fig – 4.9: Combination of s - orbitals

Combination of P - orbitals:

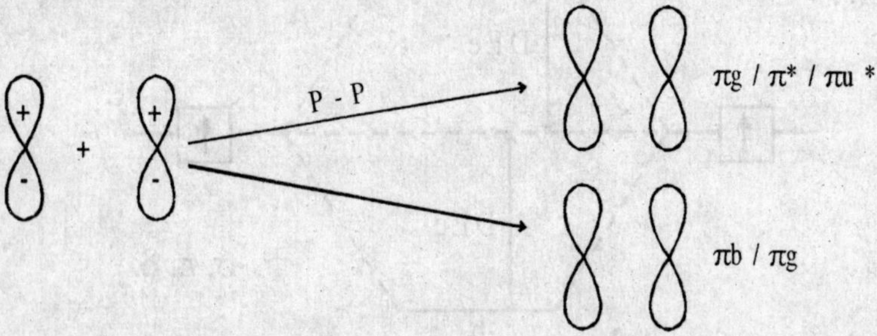

Fig 4.10 : Combination parallel to the axis.

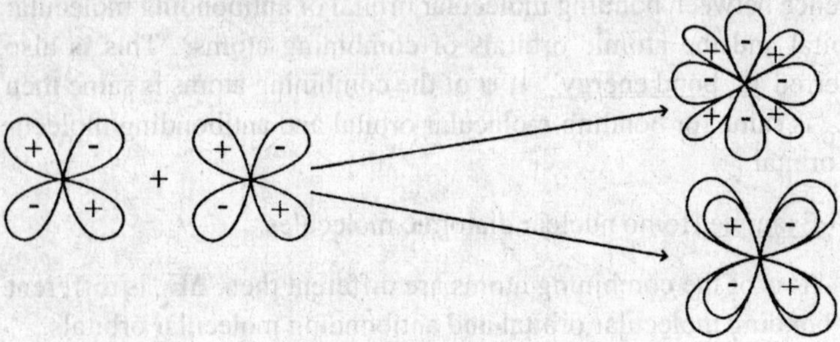

Fig 4.11 : Combination along various planes.

dxy - xy planes - lobes in between x and y axis

dyz - yz planes - lobes in between y and z axis

dy^2 - xz planes - lobes in between x and z axis

dx^2 - xz planes - lobes in between x and z axis

dx^2 - xz - in xy planes - lobes along x and z axis

dz^2 - combination of dz^2 x^2, and z^2 - y^2

Molecular orbital theory considerations:

1) In linear combination of atomic orbitals, the molecular orbit-

als are produced by combining the atomic orbitals of combining atoms.

2) Molecular orbitals do not belong to one single atom, but belongs to all the atoms and it is referred as polycentric molecule.

3) Occupation of electrons into molecular orbitals will be in the order of increasing energies

4) Total number of molecular orbitals will be equal to the total number of atomic oribitals of combining atoms.

5) The placement of electrons in the molecular orbitals follow the Pauli's principle and Hund's rule of maximum multiplicity.

6) The energies of the combining orbitals should be comparable. Otherwise overlapping is not possible.

7) The overlapping of atomic orbitals gives rise to bonding molecular orbitals, antibonding molecular orbitals and nonbonding molecular orbitals.

8) For the formation of bonding orbitals.

   i) overlap between the atomic orbitals must be +ve

   ii) Energies of the two atomic orbitals should be nearly the same.

Molecular Orbital theory

In Mo approach three centre four electron bond is assumed but the results are that covalent bond is smeared overall the three atoms. In symmetric hydrogen bonds it is equal on both sides, in unsymmetric hydrogen bonds more electron density is concentrated in the shorter link.

The MO theory considers the formation of 3 – centered MOs from the combination of AOs atomic orbitals of the three atoms – H atom and the two electronegative atoms A and B. 1s orbital of H atom, 2p orbs of A and B atoms combine to form three MOs – a bonding Mo, a non bonding MO and an anti bonding MO. Bonding MO has the lowest energy, anti bonding MO the highest energy and nonbonding MO has the same energy as that of 2p orbitals. These MOs accommodate four electrons, two from B and one each from H and H atoms. Two electrons are accommodated in bonding MO and two electrons in nonbonding MO. The bond order = $\frac{2}{3}\left(\frac{\text{no of bonding electrons}}{\text{no of atoms}}\right)$

For symmetric atoms A and B, these MOs are symmetrical but for the asymmetric bonded atoms, the MOs are such that the bonding MO extends more towards the more electronegative atom.

1) Molecular orbital energy level diagram of molecules

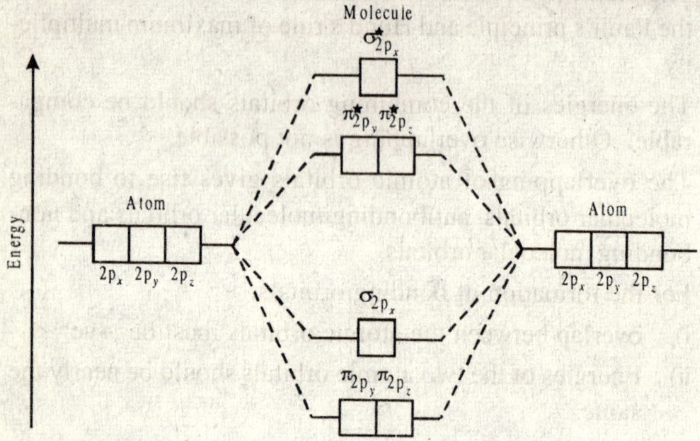

Fig 4.12: General molecular orbital energy level diagram of molecules

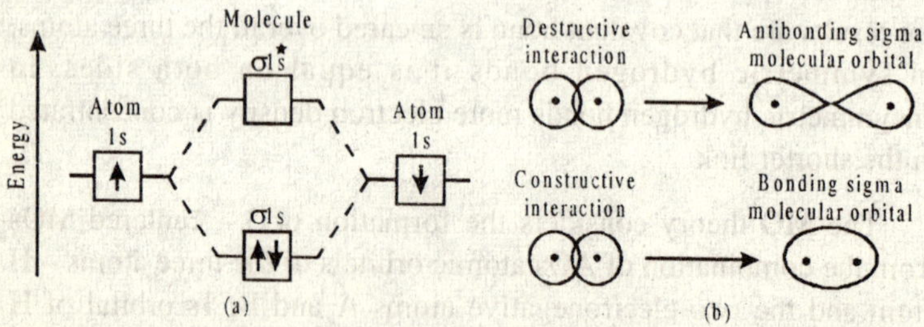

Fig 4.13: General molecular orbital energy level diagram of hydrogen molecule

$$\text{Bond order} = \frac{\text{Number of } e^- \text{ in. bonding MOs} - \text{Number of } e^- \text{ in antibonding mo's}}{\text{Number of Combining atoms.}} = \frac{2-0}{2} = 1$$

The H - H bond is perfect covalent bond. Bond distance = 4.1 pm and bond energy = -458 kJ/mol. Bond order = 1, only one bond between two H_2 atoms is formed which is a 'σ' bond.

2) Helium molecule ($1s^2$)

Helium molecule ($1s^2$)

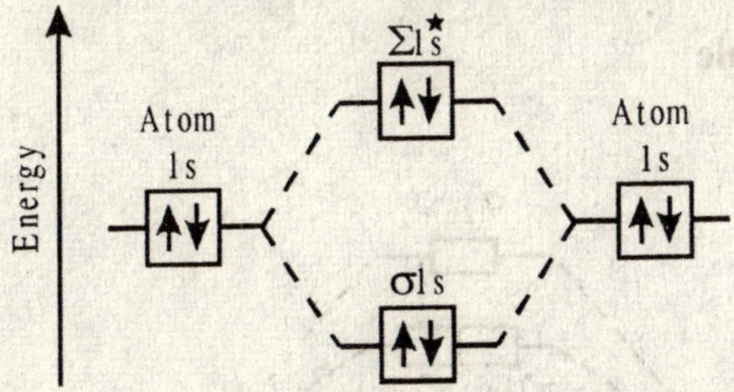

$$\text{Bond order} = \frac{\text{Number of } e^- \text{ in. bonding MOs} - \text{Number of } e^- \text{ in antibonding mo's}}{\text{Number of Combining atoms.}} = \frac{2-2}{2} = 0$$

∴ He_2 molecule does not exist.

3) Lithium molecule

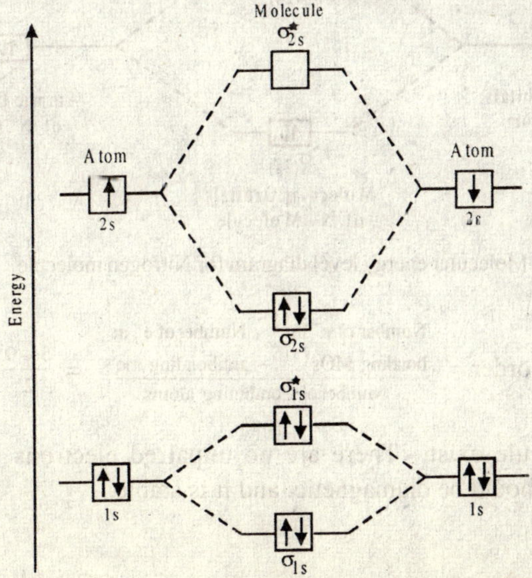

Fig 4.14: Molecular energy level diagram for Lithium molecule.

$$\text{Bond order} = \frac{\text{Number of } e^- \text{ in. bonding MOs} - \text{Number of } e^- \text{ in antibonding mo's}}{\text{Number of Combining atoms.}} = \frac{2-2}{2} = 1$$

$\therefore$ Li_2 molecule is stable if it forms. Li_2 exist in vapour state.

4) Nitrogen molecule

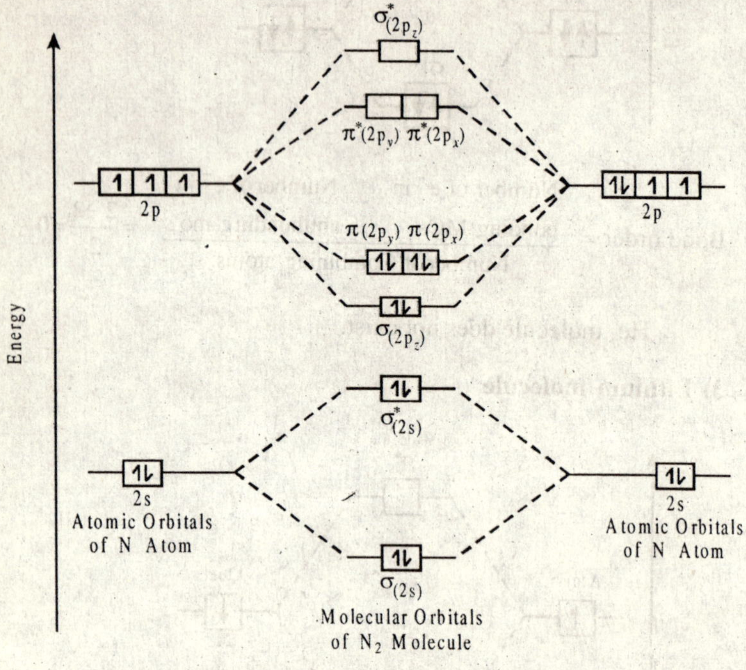

Fig 4.15: Molecular energy level diagram for Nitrogen molecule.

$$\text{Bond order} = \frac{\text{Number of } e^- \text{ in. bonding MOs} - \text{Number of } e^- \text{ in antibonding mo's}}{\text{Number of Combining atoms.}} = \frac{6-0}{2} = 3\,.$$

N_2 molecule exist. There are no unpaired electrons in the molecule. It should be diamagnetic. and it is stable.

5) O_2 - molecule: **$O_2 - 1s^2\ 2s^2\ 2p^4$.**

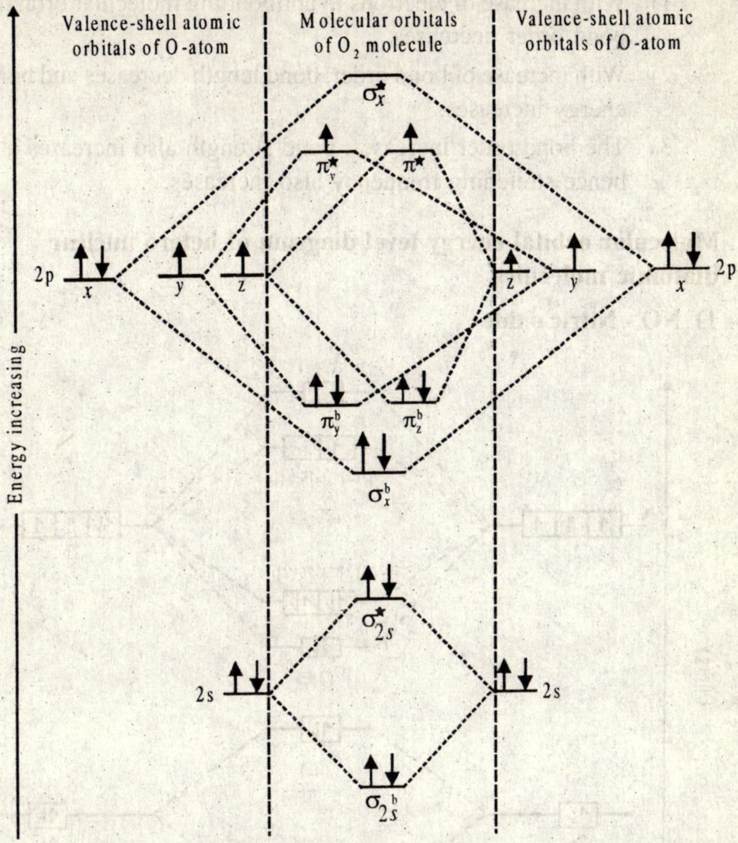

Fig 4.16: Molecular energy level diagram for Oxygen molecule.

$$\text{Bond order} = \frac{\text{Number of } e^- \text{ in bonding MOs} - \text{Number of } e^- \text{ in antibonding mo's}}{\text{Number of Combining atoms.}} = \frac{6-2}{2} = 2$$

Since there are two unpaired electrons in antibonding molecular orbitals, O_2 molecule is paramagnetic in nature.

Thus molecular orbital theory satisfactorily account for the magnetic property of O_2 molecule where as valence bond theory fails to account for this.

From the above examples. It is observed that,

1) With increase of electrons in antibonding molecular orbitals, bond order decreases.
2) With increase of bond order, bond length decreases and bond energy increases.
3) The bond order increases, bond strength also increases and hence stretching frequency also increases.

Molecular orbital energy level diagram of hetero nuclear diatomic molecules

1) NO - Nitric oxide

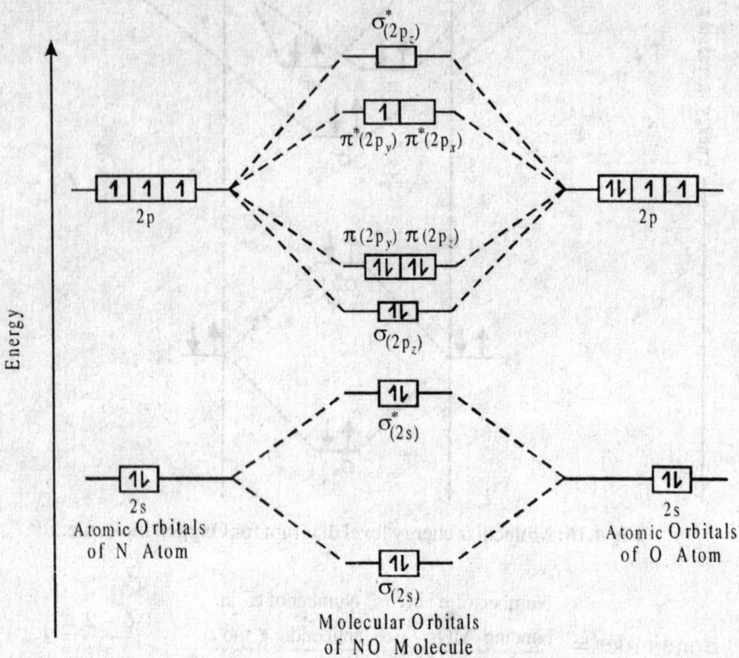

Fig 4.17: Molecular energy level diagram for NO (Nitric oxide) molecule.

$$\text{Bond order} = \frac{\text{Number of } e^- \text{ in. bonding MOs} - \text{Number of } e^- \text{ in antibonding mo's}}{\text{Number of Combining atoms.}} = \frac{6-1}{2} = 2.5$$

The 2s orbitals of O_2 and that of 'N' do not have comparable energies. They do not interact. The 2s orbital of 'O_2' remains same and has same energy.

If we consider the interaction of 2s orbital of O_2, we cannot account for the bond order and observed properties of NO molecule. Bond energy is given by 6.18 kJ/mole. NO^+ and NO^- are diamagnetic in nature and it is toxic.

Molecule	Bond order	Bond length	Bond energy	
1 NO^+	$\frac{6-0}{2}=3$	106pm	higher	>1876
2 NO	$\frac{6-1}{2}=2.5$	115 pm	6.18	1876
3 NO^-	$\frac{6-2}{2}=2$	124pm	lower	<1876

2) Carbon monoxide - CO:

C – $1s^2\ 2s^2\ 2p^2$

O – $1s^2\ 2s^2\ 2p^4$

2s orbitals of O_2 and that of carbon do not interact.

$$\text{Bond order} = \frac{\begin{matrix}\text{Number of } e^- \text{ in.} & \text{Number of } e^- \text{ in} \\ \text{bonding MOs} & - \text{ antibonding mo's}\end{matrix}}{\text{Number of Combining atoms.}} = \frac{6-0}{2} = 3$$

Since lone pair of electron are present in higher non-bonding molecular orbitals, mostly belong to atom and hence can be easily donated towards the metal ion. Hence in CO, carbon atom is a donor atom but not oxygen atom and hence 'CO' is a Ligand. It can easily donates lone pair of electrons. This diagram account for the donor property of 'CO' as a ligand.

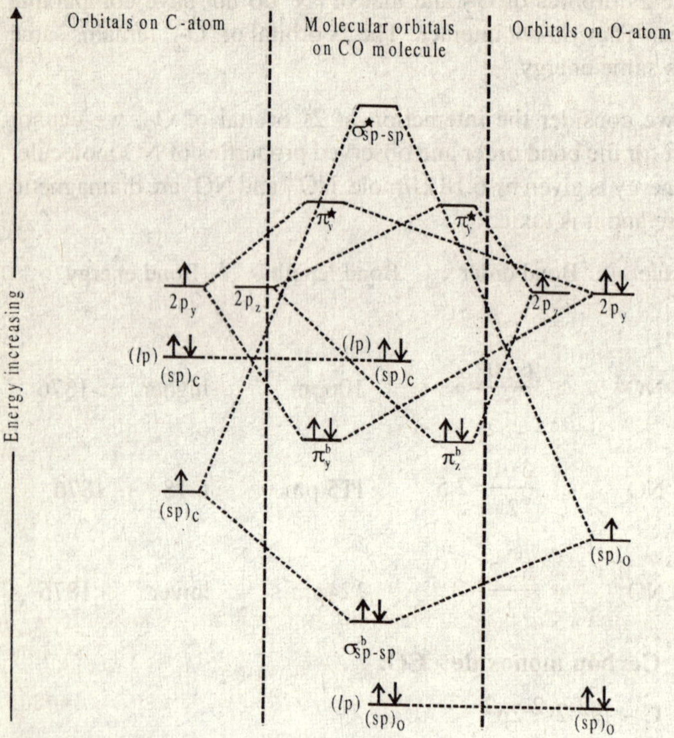

Fig 4.18: Molecular energy level diagram for Carbon monoxide molecule.

Bond energy

Bond energy is a measure of stability of a molecule. It is defined as the enthalpy change required to break a particular bond in one mole of gaseous molecules. The experimentally determined bond energy of the diatomic hydrogen molecule is

$$H_2(g) \rightarrow H(g) + H(g) \qquad \Delta H^0 = 436.4 \text{ KJ.}$$

This equation indicates that breaking a covalent bonds in one mole of gaseous H_2 molecules requires 436.4 kJ of energy.

Example : (1) $Cl_2(g) \rightarrow Cl(g) + Cl(g) \quad \Delta H^0 = 242.7 \text{ KJ.}$

Bond energy of chlorine is less than hydrogen therefore the Cl_2 molecule is less stable.

(2) Bond energies can also be calculated for diatomic molecules containing unlike elements like HCl.

$HCl(g) \rightarrow H(g) + Cl(g)$ $\Delta H = 431.9$ kJ.

(3) The bond energies of double or triple bond containing atoms can also be calculated.

$O_2(g) \rightarrow O(g) + O(g)$ $\Delta H = 498.7$ kJ.

$N_2(g) \rightarrow N(g) + N(g)$ $\Delta H = 941.4$ kJ.

It can be concluded that the bond energy of triple bond is more than double bond, which is more than single bond.

Measuring the strength of a polyatomic molecules is complicated.

Bond Length

A bond is formed between any two atoms by overlapping of atomic orbitals. This overlapping of atomic orbitals to form molecular orbitals occurs only at certain distances between the atoms. This distance between the nuclei is called bond distance or bond length. The bond distance between two given atoms is constant in different molecules. For example C – H bond length is same in most of the hydrocarbons. The normal covalent bond lengths are given as follows.

C – C bond	154 Pm.
C – N bond	147 Pm.
C – O bond	143 Pm.
C – F bond	142 Pm.
C – Cl bond	177 Pm.
C – Br bond	191 Pm.
C = C	140 Pm.
C ≡ C	121 Pm.
C = O	120 Pm.

Covalent radius of an atom A is half of the bond distance of A – A molecule. Knowing the bond length of A – A, the covalent radii of 'A' can be calculated.

For hetero nuclear diatomic molecule like A – B. Bond length of A – B = $r_A + r_B$.

For example: Bond length of HCl can be calculated as follows.

Bond length of H – H = 74 Pm

$$\text{radius of H} = \frac{74}{2} = 37 \text{ Pm.}$$

Bond length of Cl – Cl = 198 Pm

$$\text{radius of Cl} = \frac{198}{2} = 99 \text{ Pm}$$

$$\therefore \text{Bond length of H – Cl} = r_H + r_{Cl}$$

$$= 37 + 99$$

$$= 136 \text{ Pm}$$

Bond Angle

The angle between the bonding electron pairs in a given molecule or an ion is called bond angle. For example H – C – H bond angle in CH_4 molecule is equal to 109.5^0.

Factors that affect the magnitude of bond angle

1. Number of lone pair of electron on the central atom of a molecule.

The bond angle in CH_4, NH_3, and H_2O are explained as follows.

Due to the absence of lone pair of electrons as C – atom in CH_4 molecule H – C – H bond angle in this molecule is equal to the expected tetrahedral angle (109.5^0). In ammonia molecule N – atom has one lone pair of electrons on it. Here lone pair – bonded pair repulsion makes three bonding electron pairs come closer to each other and hence H – N – H bond angle is decreased to 107.5^0 instead of 109.5^0.

In H_2O molecule, O atom has two lone pairs of electrons, and two bonding pair of electrons. Here the repulsion between the different types of electron pair decreases in the order lone pair – lone pair > lone pair – bonded pair > bonded pair – bonded pair. Therefore the bond angle in H_2O molecule is less compared to NH_3, i.e., 104.5^0.

The bond angles of the above three molecules can be written as $CH_4 > NH_3 > H_2O$.

Class of molecule	Total number of electron pairs	Number of bonding pairs	Number of lone pairs	Arrangement of electron pairs*	Geometry	Examples
AB_2E	3	2	1	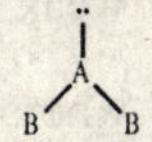 Trigonal planar	Bent	SO_2, O_3
AB_3E	4	3	1	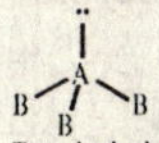 Tetrahedral	Trigonal pyramidal	NH_3
AB_2E_2	4	3	1	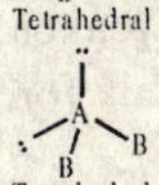 Tetrahedral	Bent	H_2O
AB_4E	5	4	1	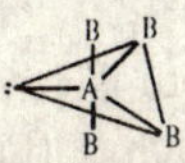 Trigonal bipyramidal	Distorted tetrahedron (or seesaw)	IF_4^+, SF_4, XeO_2F_2
AB_3E_2	5	3	2	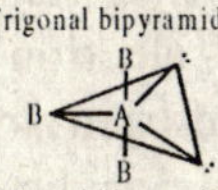 Trigonal bipyramidal	T-shaped	ClF_3
AB_2E_3	5	2	3	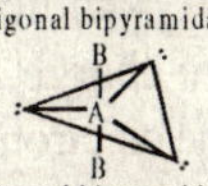 Trigonal bipyramidal	Linear	XeF_2, I_3^-
AB_5E	6	5	1	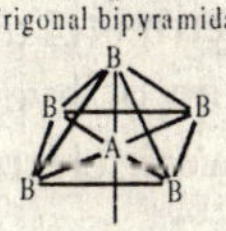 Octahedral	Square pyramidal	BrF_5, $XeOF_4$
AB_4E_2	6	4	2	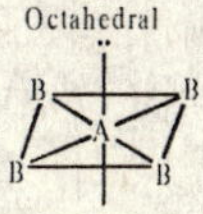 Octahedral	Square planar	XeF_4, ICl_4^-

*The colored lines are used to show the overall shape, not bonds.

Fig. 4.19: Geometry of molecules and ions in which the central atom has one or more long pairs.

2. Type of hybridization in central atom of a compound.

sp hybridization leads to the linear arrangement of molecules hence the bond angle is 180^0. sp^2 hybridization leads to trigonal or pyramidal structure hence bond angle is 120^0. sp^3 hybridization leads to tetrahedral structure where the bond angle is 109.5^0.

HYDROGEN BOND

A hydrogen bond is said to be formed when hydrogen atom is bonded to two electronegative atoms, to one by a strong covalent bond and to the other by weak electrostatic forces.

A – H - - - - - - - - B where A and B are electronegative atoms. They may be same or different. The dotted line represents hydrogen bond.

Hydrogen bond in weaker than covalent and ionic bonds but stronger than Vander waals forces.

When hydrogen atom is covalently bonded to small and highly electronegative atoms such as fluorine, oxygen and nitrogen. Due to unequal distribution of electron pair forming the covalent bond between A and H, the hydrogen atom acquires a small positive charge (thus H becomes slightly acidic) and the electronegative atom A acquires a small negative charge. The hydrogen atom carrying small positive charge attracts the electronegative atom B with negative charge or with a lone pair of electrons of another molecule or same molecule resulting in the formation of hydrogen bond. Hydrogen bond is a weak bond formed by weak electrostatic force.

$$\text{A - H} \qquad :\text{B} \qquad \overset{\delta-}{\text{A}} - \overset{\delta+}{\text{H}} \text{ - - - - - - - - - - } \overset{\times}{\underset{\times}{}}\text{B}$$

↓ Hydrogen bond

Hydrogen bond in A – H . . . . B can be either linear or significantly nonlinear.

Hydrogen atom is attached to atom A by a normal covalent bond and attached to atom B by a longer, weak hydrogen bond. This type of hydrogen bond is termed as unsymmetric.

But in some cases such as HF_2^- ion (FHF^-) (and the carboxylate) the hydrogen atom is almost intermediate between the two fluorine atoms. This type of hydrogen bonding is termed as symmetric. F – H . . . . F^-

Strong hydrogen bonds are formed when the atom A is F, O or N; weaker hydrogen bonds are sometimes formed when atom A is C or a 2nd row elements like P, S, Cl or even Br. Strong hydrogen bonds are formed when the atom B is F, O or N; the other halogens Cl, Br are less effective unless negatively charged and the atoms C, S and P can also act as B in weak hydrogen bonds. Higher the electronegativity of the atom, stronger will be the hydrogen bond. Strength of the hydrogen bonding varies from 10 – 45 kJ

Though N and Cl have the same electronegativity, N has considerably greater hydrogen bonding power. This may be due to the small size of nitrogen atom relative to chlorine atom, as a result of which the electrostatic interactions are stronger than those of Cl.

When hydrogen bond is formed, the Vander waals distance Between H and B atoms in the system A – H - - - - - B is found to be greater than the observed distance. This indicates that penetration of electron cloud of B by the H atom. This contraction of inter ionic distances which can be detected from X – ray or neutron diffraction studies is an evidence for the existence of hydrogen bonds.

Calculated Vander waals distances and observed distances.

A – H - - - - - - B	H - - - - - -B	H- - - - - - -B	A- - - - - - -B	A- - - - - -B
Bond	Calculated	Observed	Calculated	Observed
F – H - - - - - -F	260 Pm	120 Pm	270 Pm	240 Pm
O – H - - - - - - O	260 Pm	170 Pm	280 Pm	270 Pm

Two types of hydrogen bonds exist in compounds

1. Intermolecular hydrogen bonding
2. Intermolecular hydrogen bonding

1. When the hydrogen bond is formed between molecules then, it is known as Intermolecular hydrogen bonding.

Eg., HF, H_2O, NH_3, HCOOH, CH_3COOH etc.,

Intermolecular hydrogen bonding leads to association of molecules.

1. Hydrofluoric acid – HF: Due to intermolecular hydrogen bonding a number of HF molecules get associated through hydrogen bonding in Zig – Zag chains.

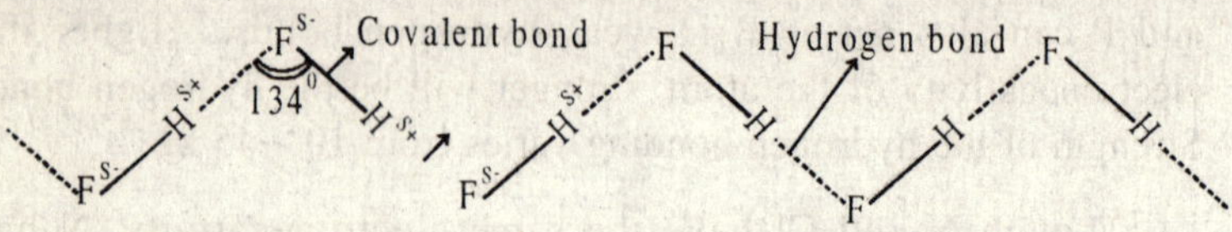

2. Water: Water molecules are associated through hydrogen bonding as follows

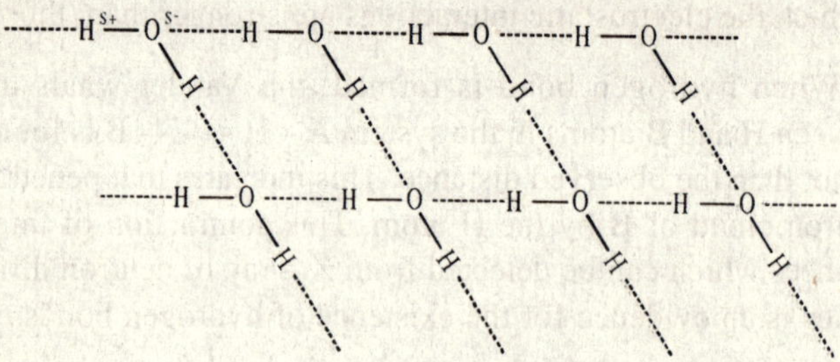

3. Ammonia: Similarly in ammonia molecules, a number of NH_3 molecules are associated through hydrogen bonding.

4. Formic acid: Two molecules of HCOOH are associated through hydrogen bonding to form a dimer.

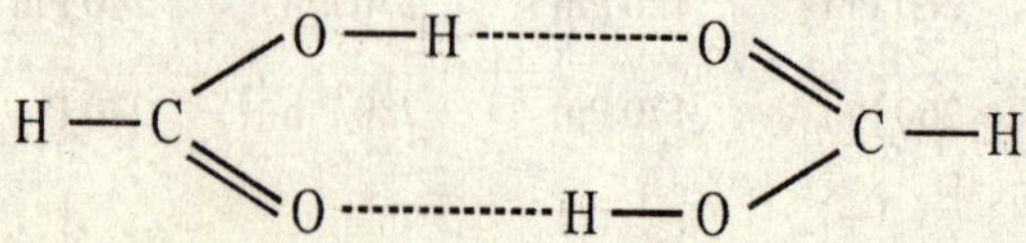

When hydrogen bond is formed within the molecule, then it is known as Intra molecular hydrogen bonding.

Eg.: O – nitro phenol, O – amino phenol, salicylaldehyde, Pyridine – 2 – carboxaldoxime.

Intra molecular hydrogen bonding leads to chelation (i.e., ring formation).

Necessary conditions for the formation of intra molecular hydrogen bonding are

a) Ring formed as a result of hydrogen bonding should be planar.
b) A five or six membered ring should be formed.
c) Interacting atoms should be placed in such a way that there is minimum strain during the ring closure.

O – nitro phenol: Hydrogen bond is formed Between hydrogen atom of OH group and one of the oxygen atoms of NO_2 group within the molecule.

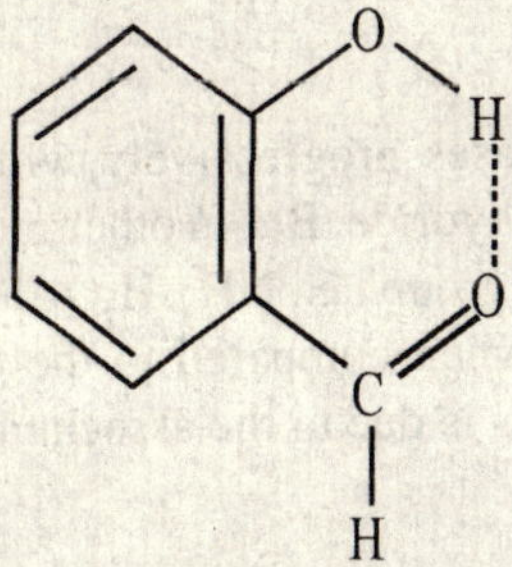

Intra molecular hydrogen bonding is possible in O – nitro phenol where a six membered ring is formed but it is not possible in m - and P – nitro phenols because of the larger size of the ring which would have to be formed. In m and P – nitro phenols intermolecular hydrogen bonding takes place.

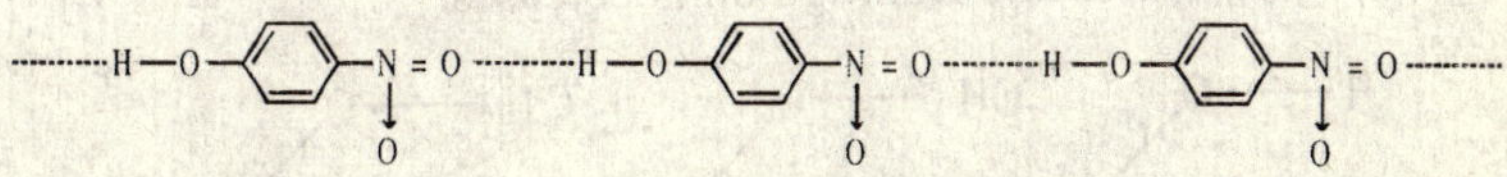

Salicylaldehyde

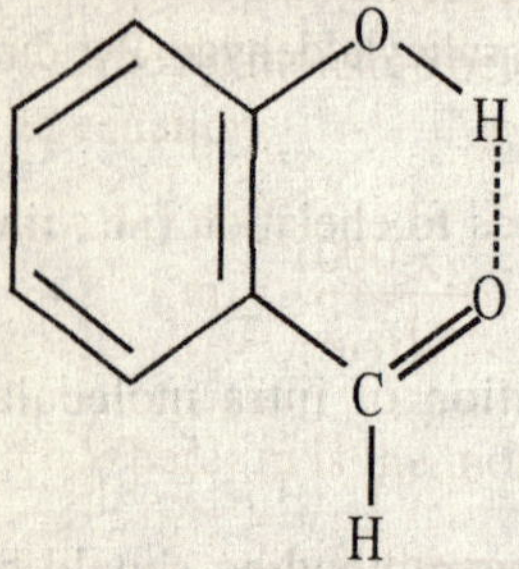

O-amino phenol

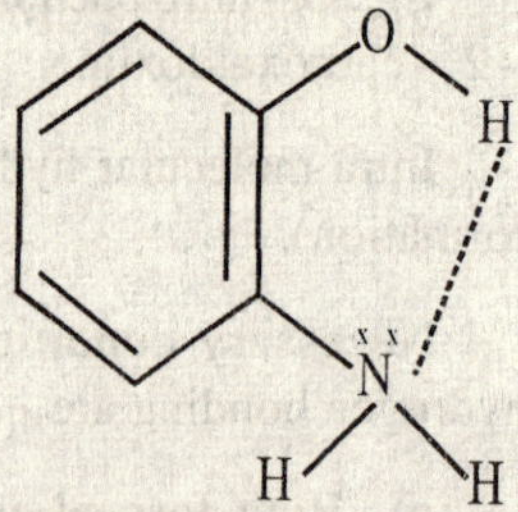

Though a weak bond, hydrogen bond affects the physical properties of compounds

Example :

1. Boiling point of liquids: Intermolecular hydrogen bonding leads to increase in boiling point of liquids. This is due to the association of molecules through hydrogen bonding.

Consider the boiling points of hydrides of groups 14, 15, 16 and 17 elements.

In group 14, Boiling Point decreases progressively with the decrease in the molecular weight of the hydride. But in other groups. The hydride of the first member of each group i.e., NH_3, H_2O and HF have abnormally high boiling points when compared to the other hydrides of their respective groups. This is due to the association of molecules through hydrogen bonding.

For the same reason, NH_3, H_2O and HF are liquids at room temperature whereas hydrides of other elements of group 15, 16 and 17 are gases.

In H_2O, CH_3OH, CH_3OCH_3, as the hydrogen atoms in H_2O are replaced progressively by the - CH_3 groups, the extent of hydrogen bonding decreases and Boiling Point decreases.

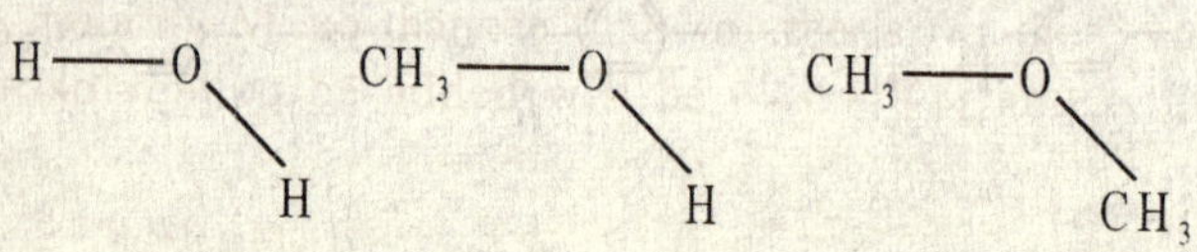

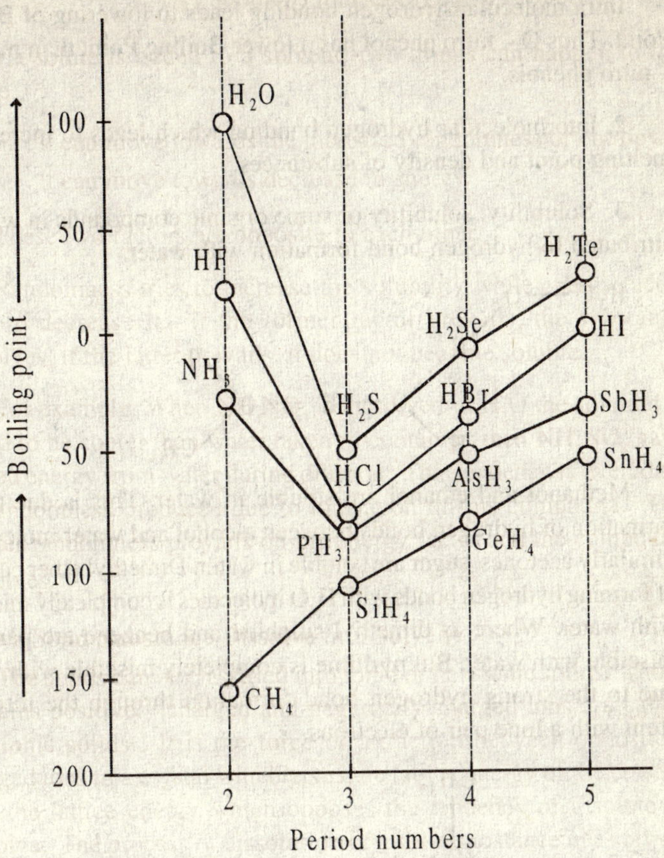

	gr 14	gr 15	gr 16	gr 17
	CH_4	NH_4	II_2O	HF
BP	SiH_4	PH	H_2S	HCl
	GeH_4	AsH_3	H_2Se	HBr
	SnH_4	SbH_3	H_2Te	HI

Intra molecular hydrogen bonding leads to lowering of Boiling Point. Thus O – nitro phenol has a lower Boiling Point than m and P – nitro phenols.

2. Intermolecular hydrogen bonding which leads to increase in melting point and density of substances.

3. Solubility: solubility of some organic compounds in water is attributed to hydrogen bond formation with water.

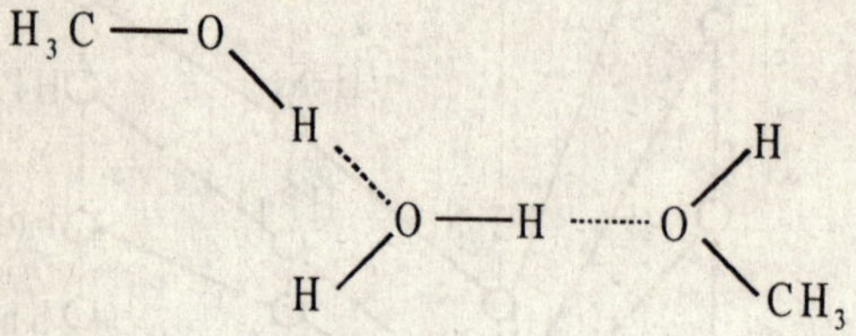

Methanol and ethanol are soluble in water. This is due to the formation of hydrogen bonds between alcohol and water molecules. Similarly acetones, sugar are soluble in water. Dimethyl ether capable of forming hydrogen bonds with H_2O molecules is completely miscible with water. Where as dimethyl sulphide and benzene are partially miscible with water. But pyridine is completely miscible with water due to the strong hydrogen bond formation through the nitrogen atom with a lone pair of electrons.

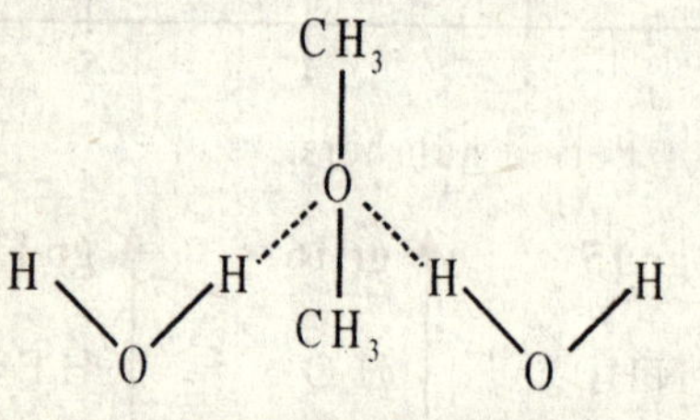

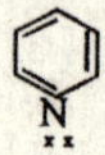

4. Viscosity: Intermolecular hydrogen bonding leads to increase in viscosity of liquids.

Intermolecular hydrogen bonding increases attractions between the molecules in the different layers of the hydrogen bonded liquids. Due to hindrance in the smooth flow of the liquid layers over one another, viscosity increases.

5. Azeotropic mixtures: There are many liquids which have no hydrogen bonding but have slightly positively charged hydrogen atom. Eg., halogenated Hydro Carbons like $CHCl_3$

There are many other liquids which are also not hydrogen bonded but have electron donor atoms

Eg., Pyridine, Ketones, esters, Carboxylic acids etc.,

If the liquids of two classes are mixed, intermolecular hydrogen bonding is formed between them such mixtures form high azeotropic mixtures

Eg., Mixtures of $CHCl_3$ – acetone, $CHCl_3$ – Pyridine etc.,

6. Dielectric constant; Intermolecular hydrogen bonding leads to increase in dielectric constant. This is due to the formation of polymeric species having enhanced dipole moments rather than sum of the dipole moments of constituent molecules.

7. Anomalous properties of water: Density of water is higher than that of ice and water has a maximum density of 1.00 g cm^{-3} at 4^0C

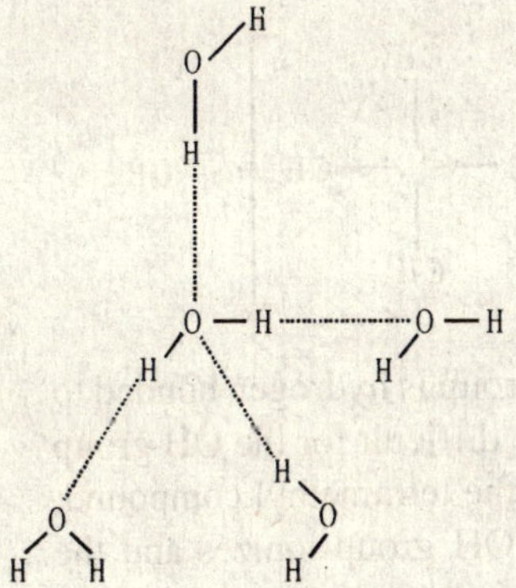

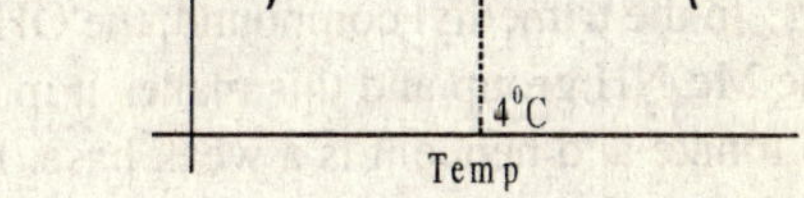

In ice tetrahedral arrangement of oxygen atoms around each oxygen atom is formed. Due to this arrangement, ice has large open structure, large volume and low density. When ice melts, the open structure breaks down to give more compact water phase having high density. (Some of the hydrogen bonds are broken and new hydrogen bonds are formed so that H_2O molecules come closer to each other) And hence density of water is higher than that of ice. Breaking and reformation of hydrogen bonds is completed at 4⁰C and at 4⁰C water has a density of 1.00 g cm^{-3}. With increase in temperature, hydrogen bonds are further broken but no new hydrogen bonds are formed, H_2O molecules are separated, volume increase and density decreases with increase in temperature, therefore water has a maximum density of 1.00 g cm^{-3} at 4⁰C.

Some important properties of hydrogen bonding

Hydrogen bonds are responsible for water being liquid at room temperature but for this life would exist on water.

Since hydrogen bonds have low bond energy, they also have a low activation energy and this results in their playing an important part in many reactions at normal temperature.

Hydrogen bonding may be used to explain the strength of bases.

Eg.,

Trimethyl ammonium hydroxide

$(CH_3)_3$ NHOH or $(CH_3)_3$ NH^+OH^-

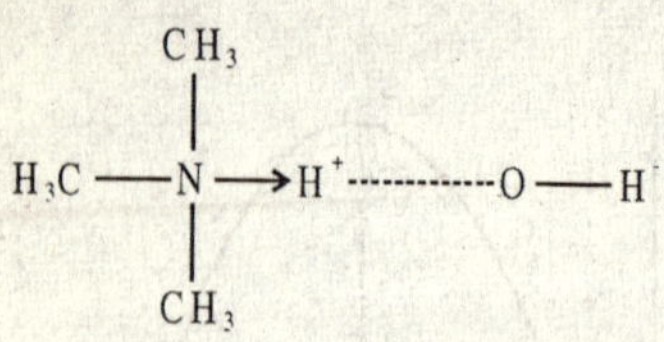

Tetramethyl ammonium hydroxide

$(CH_3)_4$ NOH or $(CH_3)_4$ N^+ OH^-

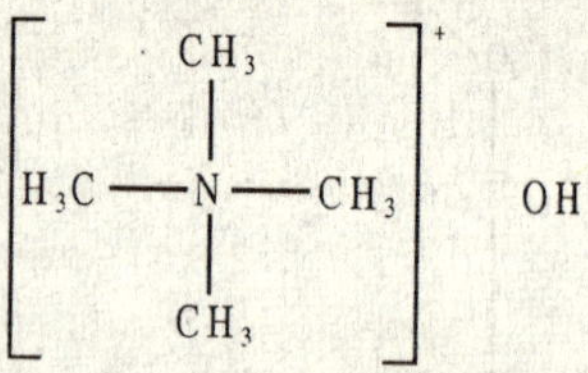

In the trimethyl compound, the OH group is hydrogen bonded to the Me_3NH group and this makes it more difficult for the OH group to ionize and hence it is a weak base. In the tetramethyl compound, hydrogen bonding cannot occur, so the OH group ionizes and the compound is thus a strong base.

Despite low bond energy hydrogen bonds are of great importance both in biochemical systems and in normal chemistry, hydrogen bonds are extremely important. For example: they are responsible for linking pairs of bases in large nucleic acid containing molecules and for linking polypeptide chains in proteins. The hydrogen bonds maintain these large molecules in specific molecular configurations which is important in the operation of genes and enzymes.

Structure of Proteins: Proteins are polypeptides consisting of long chain of amino acid residues linked by the peptide linkages (-NH – CO – bond).

(The peptide links marked by thicker lines have double bond character and so are resistant to rotation)

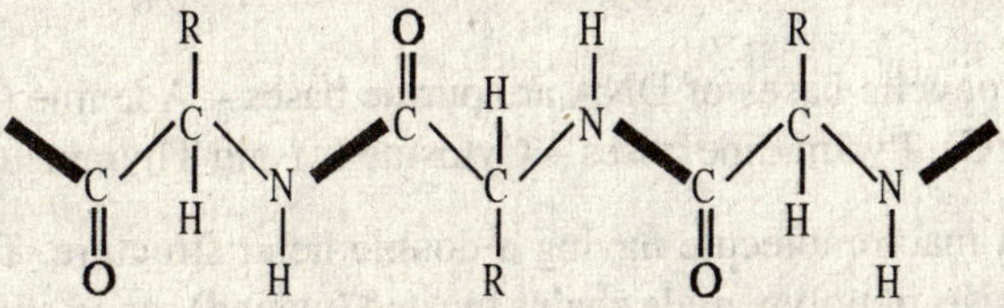

Rotation is not possible about – NH – CO bond but possible about C – C bond and C – N bond thus the long chain of protein molecules, is coiled (twisted) to form a helix. Hydrogen bonding occurs between a N-H group and a C = O group of the same chain. These hydrogen bonds hold the helix together and the molecule has a specific molecular configuration (Each amide group is attached by a hydrogen bond to the third amide group in it in both directions along the chain resulting in an ∝ - helix)

Structure of nucleic acids: Nucleic acids DNA and RNA consist of Poly nucleotide chains. A nucleotide consists of a residue pentose sugar (Deoxy ribose or ribose), one of a purine or a pyrimidine base and one of phosphoric acid bonded together.

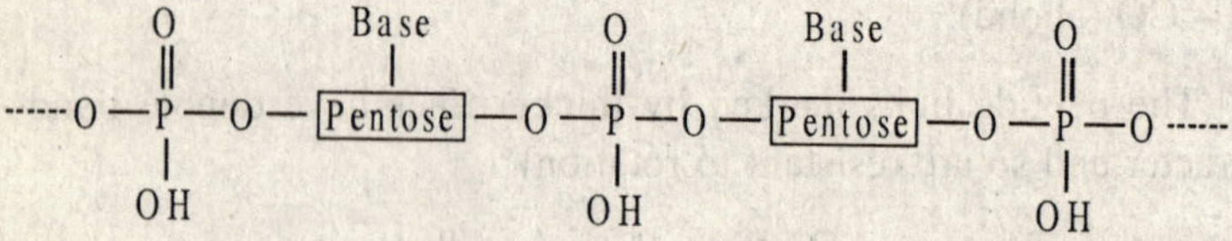

The heterocyclic bases of DNA are purine bases – Adenine (A) and Guanine (G), Pyrimidine bases – Cytosine (c) and Thymine (T)

DNA is a macromolecule having a double helix structure. The molecule consists of polypeptide chains twisted (wound) round about each other to form a double helix. The two chains are oriented in opposite directions (are anti parallel), the tail of one is opposite to the head of other. The heterocyclic bases point towards the inside of the helix. The two chains are held together by hydrogen bonds formed between a purine of one chain and a pyrimidine of another chain.

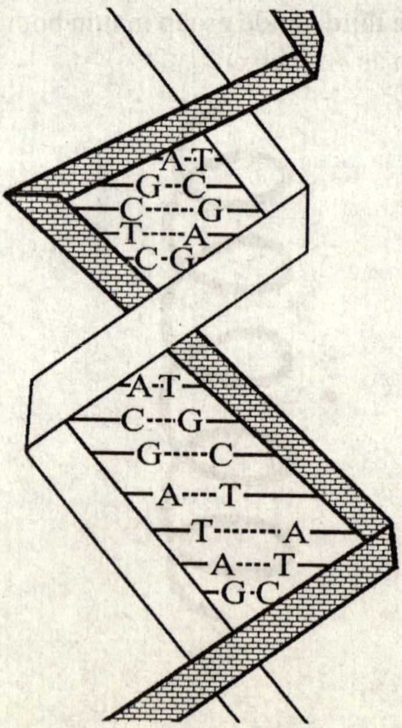

Adenine of one chain is linked to Thymine of another chain through hydrogen bonding. Similarly, Guanine of one chain is linked to cytosine of another chain through hydrogen bonding.

– A ....T –   – G ....C –

Because of formation of hydrogen bonds, the DNA molecule has a definite molecular configuration.

Detection of hydrogen bonding

Hydrogen bonding may be detected from X – ray diffraction studies, neutron diffraction studies, IR spectral studies, and proton N M R spectral studies. The best methods for the detection of hydrogen bonding are Far IR and Proton N M R spectral studies.

Theories of hydrogen bonding

Some of the common theories advocated for the formation and stability of hydrogen bonds are

Dipole – Dipole or Dipole – Ion interaction theory

This theory considers the electrostatic force of attraction when hydrogen bond is formed with electronegative atoms. The interaction may be dipole – dipole interaction or dipole –ion interaction.

$A^{\delta-} - H^{\delta+}$ .... $B^{\delta-} - R^{\delta+}$ or $A^{\delta-} - H^{\delta+}$ .... $B^{\delta-}$

(Where R is the remainder of the molecules containing the electronegative atom)

This view was supported by the fact that

Strongest hydrogen bonds are formed in systems in which hydrogen is bonded to the most electronegative atoms.

Bond	*Energy of H bond*	*The bond energy of hydrogen bond*
F – H......F	23 – 25 kJ mol^{-1}	is HF_2^- ion (eg...KHF_2) is the
O – H......O	20 – 30 kJ mol^{-1}	highest and is equal to 113.4 kJ mol^{-1}

b) The bonded atoms tend to be linear (with some exceptions) due to electrostatic repulsions.

But the theory does not explain

Shortening of bond distance in H . . . . . .B

The repulsion between the nucleus of H and B at short distances.

Symmetry of HF_2^- ion where H atom is intermediate between the two fluorines

$$\overset{\delta+}{F-H}\ldots\ldots\overset{\delta-}{F^-} \leftrightarrow \overset{\delta-}{F}\ldots\ldots\overset{\delta+}{+1-F}$$

Resonance theory

Resonance theory considers the delocalization of covalent bond over both sides of the hydrogen atom. The hydrogen bond is taken be the resonance hybrid of the resonance structures I to VI

$$\underset{\text{I}}{A-H:B} \leftrightarrow \underset{\text{II}}{A^-H^+:B} \leftrightarrow \underset{\text{III}}{A^-H-B^+}$$

vander Waals forces:

Noble gases generally do not form any covalent bonds. Therefore, no chemical forces are operating between the atoms in the crystals of noble gases. vander Waals forces are the only attractive forces in these cases. vander Waals forces are the weakest forces in the nature. The distance of the closest approach of the two adjacent atoms of the noble gas concerned in the solid state is referred to as Vander waals distance. Half the Vander waals distance is the vander Waals radius. The Vander waals radius itself is taken as the atomic radius of the atom noble gas.

vander Waals forces are very short lived electrostatic attractive forces which exist between all kinds of atoms, molecules and ions when they are brought close to each other. vander Waals forces are not concerned with the valence electrons of the species, they exist between atoms, molecules and ions. Even the inert gas molecules do possess these forces.

Factors favouring the formation of vander Waals forces.

The factors affecting the strength and magnitude of force are as follows:

1. **Temperature:** On increasing the temperature of the system, the kinetic energy increased and velocity of molecules increase leading to decreasing in Vander waals forces between them.

2. **Number of electrons present in molecules:** When molecules contain large number of electrons, then there is more polarization of electron cloud, therefore, Vander forces become more.

3. **Molecular size:** Large molecules are more easily polarizable because the size electron cloud increases. This increased molecular interactions lead to increase in consequently magnitude and strength of the vander Waals forces.

QUESTIONS

1. What do you understand by stable electronic structure?
2. Write the electronic Configuration of PCl_5 and H_2SO_4
3. What type of bonds do you expect between (a) very small cation and very large anion. (b) between the atoms of the same element. Give reasons.
4. Explain the salient features of valence bond theory.
5. Compare the properties of ionic and covalent compounds.
6. What is electro negativity? How is it used to predict type of bonding.
7. Write Lewis dot formula for (i) HOCl (ii) BF_3 (iii) NH_4
8. Describe the structures of water and ammonia molecule in reference with VSEPR Theory.
9. What are sigma and pi bonds? How are they formed?

10. What is a coordinate covalent bond? How is it different from a covalent bond.
11. Explain the structure of methane with help of hybridization.
12. Explain inter molecular and intra molecular type of hydrogen bonding with an example.
13. What is meant by an ionic bond? Explain with an examples.
14. What are polar and non polar bonds? Explain with examples.
15. What is the significance of hydrogen bonding in bio molecules. Explain with examples.
16. Describe the basic ideas of the VSEPR theory.
17. Explain the formation of a covalent bond between two chlorine atoms.
18. Define (i) ionic bond (ii) co-ordinate bond. (iii) metallic bond.
19. What is lattice energy? Explain the calculation of lattice energy using Born Haber cycle.
20. Describe SP, SP^2, SP^3 hybridization with the help of simple organic molecules.
21. Discuss with the help of M O theory, the formation of N_2, O_2, and F_2 molecules.
22. What is the state of hybridization of the Central atom of the following?
23. a) $BeCl_2$ b) BCl_3 c) CCl_4 d) PF_3
24. Write the electronic configuration of HCl molecule. Draw its energy level diagram.
25. What is an atomic orbital ? Sketch the shape of Sand P-orbitals.
26. What do you understand by overlap of atomic orbitals?
27. Explain the types of bond formed by S - S, S - P overlap.
28. Explain the formation of H_2 molecule with valence bond and molecular orbital theory.

29. What is meant by hybridization ? Write the hybridization of water molecule.
30. Why is shape of methane molecule is tetrahedral?
31. Compare valence bond theory and molecular orbital theory.
32. Explain why Sigma bond is stronger than Pi bond.

5

RADIOACTIVITY

The phenomenon of spontaneous disintegration of heavy nuclei with the emission of certain radiation is called radioactivity. The radioactivity found in naturally occurring elements is known as natural radioactivity. The substance which exhibits radioactivity is known as radioactive substance. The radioactive substance emits either an $\propto$– particle or a β - particle. γ - emission takes place along with α & β emission. Rutherford and Soddy studied radioactivity of many elements. In the radioactivity process, the radioactive element changes over to a non – radioactive element. Example: U, Th, Po etc are the naturally accuring radioactive elements. A number of elements in nature are radioactive due to the reason that in their nucleus they possess unequal number of neutrons and protons. When the $\frac{n}{p}$ ratio is equal to 1 the nucleus is stable. If it is altered then nucleus is said to be unstable and becomes radioactive. The relation between neutron to proton content of nucleus is shown below.

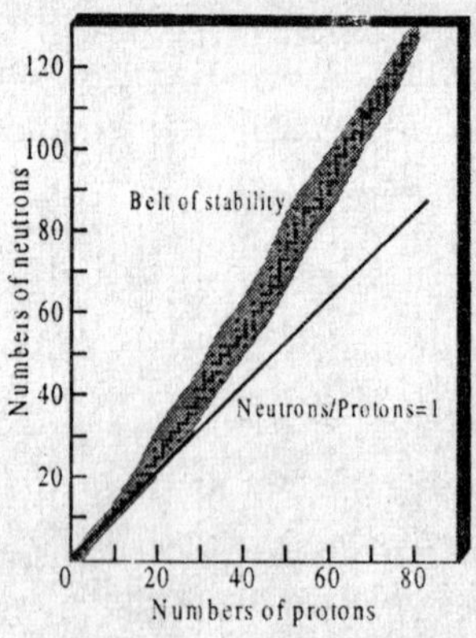

Fig. 5.1: n/p ratio of elements

Natural and Artificial radioactivity:

Naturally occurring elements like uranium, thorium, polonium, radium etc are constantly undergoing nuclear disintegration with the emission of radiation like α, β and γ - rays. This spontaneous change is called natural radioactivity. The property is found in heavy elements represented in periodic table. The natural radioactive series are classified into the three series:

1. The uranium series.
2. The thorium series.
3. The actinium series.

The uranium series is represented as follows.

The uranium decay series

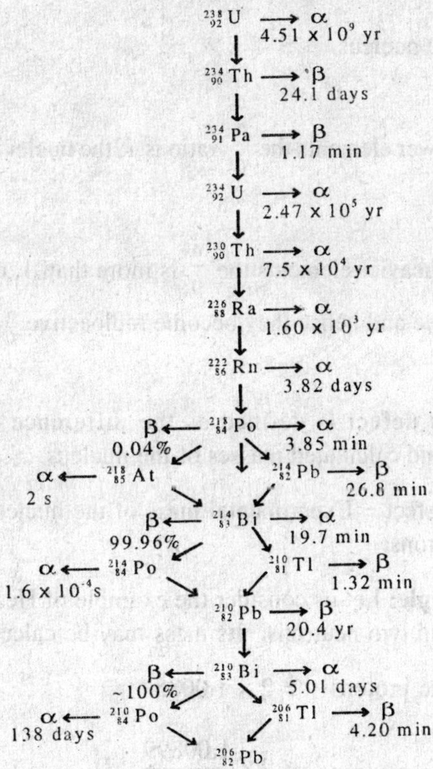

Fig.: 5.2 : The uranium series.

Artificial or induced radioactivity is a process by which a stable element is converted to a new radioactive isotope of a known element by artificial means.

Example: When α - particles are bombarded with $^{14}_{7}N$, it is converted with radioactive oxygen. $^{14}_{7}N + ^{4}_{2}He \rightarrow ^{17}_{8}O + ^{1}_{1}H$

$\frac{n}{p}$ Ratio and Nuclear Stability

The nucleus of an atom consists of protons and neutrons. The protons tend to go apart from each other as a result of the repulsive forces between them. The neutrons hold the protons together the nucleus. The stability of the nucleus depends on neutron to proton ratio $\left(\frac{n}{p}\right)$ in the nucleus.

1. For lower elements the $\frac{n}{p}$ ratio is 1, the nuclei is found to be stable.
2. For a heavier elements the $\frac{n}{p}$ is more than 1, the nucleus is unstable and hence they become radioactive.

Mass defect

The mass defect is defined as the difference between the experimental and calculated masses of the nucleus.

∴ Mass defect = Experimental mass of the nucleus - Mass of protons + neutrons.

For Example: Let us consider the example of He it consists of two protons and two neutrons. Its mass may be calculated as:

Mass of the protons = 2 × 1.0085

Mass of the neutrons $= 2 \times \frac{1.00899}{4.03248}$

However, it was found that the experimental mass of He is only 4.00388. This is less by 0.03040 amu than that calculated as above is called mass defect.

Binding Energy (BE)

The nucleus of an atom is made up of mainly protons and neutrons which are closely packed in a small volume. There are intensive repulsive forces between the component protons do exist. However they do not disturb the stability of the nucleus. This is due to the reason that the nucleus is bound to one another by powerful forces. The energy that binds the nucleus together in the nucleus is called the nuclear binding energy. According to the Einstein theory, it is the mass defect which is converted to binding energy. Hence, the binding energy is the energy equivalent to mass defect.

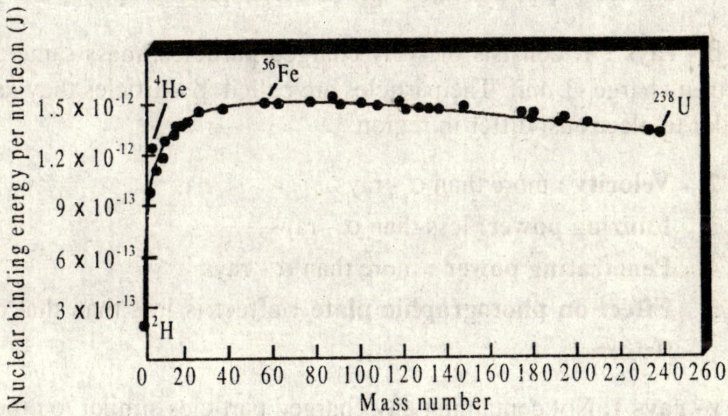

Fig. 5.3: The relationship between Mass number and Binding energy.

Calculation of binding energy

The binding energy can be calculated from the mass defect of the nucleus by using Einstein's equation.

$$\Delta E = \Delta m \ c^2.$$

Mass defect = Mass of nucleons – Mass of nucleus.

$$\text{Mass defect in grams} = \frac{\Delta m}{6.02 \times 10^{23}} \text{g since } \frac{1}{1.6604 \times 10^{-24}} = 1 \text{ amu}$$

Binding energy (BE) = mc^2

$$BE = \Delta m \times 1amu \times (3 \times 10^8)^2 \times 10^{-3} \text{ Joule}$$

$$BE = \Delta m \times amu \times 1.4924 \times 10^{-10} \text{ Joule}$$

Higher the binding energy, more stable is the nucleus.

Types of radiations :

α - rays : 1. consists of +vely charged particles, mass 4 amu, charge +2 unit. They are nothing but helium nuclei (He^{+2} 40th)

2. **Velocity:** equal to $\frac{1}{10}^{th}$ velocity of light.
3. **Ionizing power :** highest of all the rays.
4. **Penetrating Power :** least of the three rays.
5. **Effect of photographic plate:** effect photographic plate.

β - rays : 1. consists of -vely charged particles, mass same as electron charge -1 unit. The particles are called β - articles they are similar to electrons; differ in region.

2. **Velocity :** more than α - rays
3. **Ionizing power:** less than α - rays
4. **Penetrating power :** more than α - rays
5. **Effect on photographic plate :** effect is less than that of α-rays.

γ - rays 1. Not containing any charged particles similar to other electromagnetic radiation.

2. **Velocity :** same as that of light.
3. **Ionizing power:** least of all the three rays.
4. **Penetrating power :** they can penetrate through aluminum sheets of 100 cm thickness.
5. **Effect on photographic plate :** affect photographic plate.

Law of radioactivity or Soddy's group displacement law :

1. A radioactive element always emits either ∝– particle or a β - particle at a time. They may be accompanied by γ - ray.

2. When an $\propto$– particle is ejected, the new product has its atomic number decreased by 2 units and mass number decreased by 4 units.

$$_{z}X^{A} \xrightarrow{\propto} {}_{z-2}Y^{A-4}$$

3. When a β - particle is ejected, the new element formed has the atomic number increased by one unit and the mass number remains same.

$$_{z}X^{A} \xrightarrow{\beta} {}_{z+1}Y^{A}$$

4. The process of disintegration continuous until, a stable isotope of lead whose atomic number 82 is formed.

Law of radioactive decay or law of disintegration

It states that, the rate of disintegration is the number of nuclei disintegrating per second is directly proportional to the number of radioactive nuclei present in the given mass of the element at that instant.

If N is the number of atoms present and dN is the number of atoms disintegrating in time dt, then $\frac{dN}{dt} \propto -N$

Negative sign shows that the number of radioactive atoms decreases with time.

$$\therefore \frac{dN}{dt} = -\lambda N$$

λ' is a constant and is called as disintegration constant or decay constant.

$$\frac{dN}{dt} = -\lambda N \quad \text{or}$$

$$\frac{dN}{N} = -\lambda dt \quad \text{or}$$

Integrate with between limits: when t = 0; N = No; t = t ; N = N

$$\int_{N_0}^{N} \frac{dN}{N} = -\lambda \int_{0}^{t} dt$$

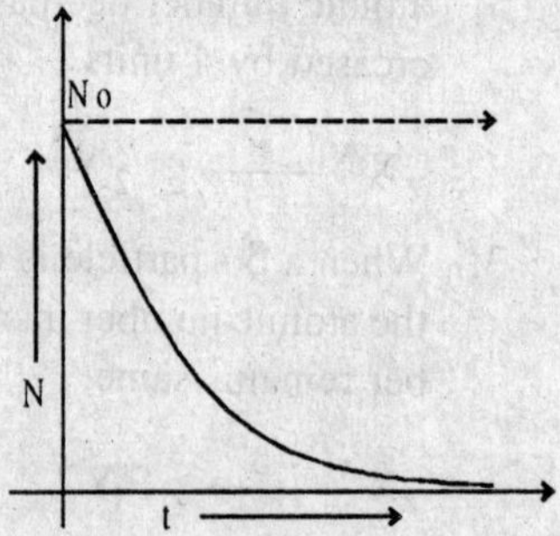

$$l[\log]^{N} \ \delta = -\lambda t \ \delta$$

$$\ln \frac{N}{N_o} = -\lambda t$$

$$N = N_o \, e^{-\lambda t}$$

This equation shows that, the radioactive atoms disintegrate exponentially with time.

Rate of radioactive decay : Consider a radio active element A disintegrating to give products

$$A \longrightarrow \text{product.}$$

Let N_0 be the number of atoms in the beginning at time t = 0, N be the number of atoms at time t, the rate of disintegration

$$\frac{-dN}{dt} \alpha N \ \frac{-dN}{N} = \lambda dt; \lambda \quad \text{disintegration constant}$$

$$\int \frac{-dN}{N} = \int \lambda.dt$$

$$- \ln N = \lambda t + c$$

When t = 0, N = No.

$$\therefore - \ln N_0 = c$$

$$\therefore - \ln N = \lambda t - \ln N_0$$

$$\ln N_0 - \ln N = \lambda t$$

$$\therefore \lambda = \frac{2.303}{t} \log \frac{N_o}{N}$$

Half life period : The time during which the activity of a radioactive substance becomes half of its original value.

When $t = t_{1/2}$ $N = \frac{N_o}{2}$

$$\therefore \lambda = \frac{2.303}{t_{1/2}} \log \frac{N_0}{\frac{N_0}{2}}$$

$$\lambda = \frac{2.303}{t_{½}} \log 2$$

$$\lambda = \frac{0.693}{t_{1/2}} \text{ or } t_{1/2} = \frac{0.693}{\lambda}$$

Mean life or average life (T) :

The mean life of radioactive element is defined as the ratio of sum of the life time of all the radioactive atoms to the number of radioactive atoms present in it. It can be shown that, $\mathcal{T} = \frac{1}{\lambda}$. The reciprocal of λ is called average life period

$$T = \frac{1}{1}; \therefore t_{1/2} = \frac{0.693}{1} = 0.693 \times T$$

$$= te^{-\lambda t} \Big|_{\infty}^{0} + \frac{e^{-\lambda t}}{+\lambda} \Big|_{\infty}^{0}$$

Law of successive disintegration and condition for radioactive equilibrium.

Statement: It states that, if a radioactive substance disintegrates to form a new substance which inturn disintegrate to form a new substance and so on, until a stable isotope of lead with atomic number 82 is formed.

Let a substance A decays to form B which inturn decays to form C and so on. Then

$A \rightarrow B \rightarrow C \rightarrow Pb.$

Unit of radioactivity:

Unit of radioactivity is curie (Ci). It is the activity of one gram of radium in which 3.7×10^{10} atoms disintegrate per second. Another unit which is commonly used is milli curie (mci). It is $\left(\frac{1}{1000}\right)^{th}$ of curie.

One Rutherford = 10^6 disintegration per second.

One Becquerel = One disintegration per second.

Rutherford unit (Rd) 1 Rd = 10^6 disintegration per second.

Theory of α - decay (or theory of α - disintegration.)

Only heavy nuclei with mass number (A) greater than 200 undergo α - decay. It must be a nuclear phenomenon. α - particles pre exists in the nucleus and are arranged in distinct energy levels. In nucleus there are short range attractive forces between the nucleons which are called nuclear force. In nuclei of heavy atoms, the short range nuclear forces are insufficient to counter balance the repulsive forces of proton. As a result, α - decay takes place and nucleus becomes stable. The binding energy of α- particle is comparatively higher than proton and neutron. Therefore α - particle emitted instead of individual proton and neutron.

Coulomb's law is applicable when α - particle is outside the nucleus. When it is inside the nucleus, the force is attractive and coulomb's law does not holds good.

Theory of b – decays:

Emission of electron (e^-) or positron (e^+) from the nucleus of radioactive substance is called β – ray emission.

There is a difficulty in expressing β - decay mechanism because in β - decay a nucleus emits an electron where as according to accepted model of atom nucleus consists of protons and neutrons. This difficulty was overcome by assuming that, β - particles do not preexists in the nucleus but are formed just the moment of transformation of neutrons into protons.

$$n \rightarrow p + \beta$$

$$\text{or} \quad n \rightarrow p + e^{\oplus}$$

Theory of γ – decay:

The emission of γ - rays, immediately follows the emission of ∝ - particles or β - particles. The emission of α particles or β - particles leaves the daughter nucleus in an exited state. The daughter nucleus comes to ground state by γ - ray emission.

$${}^{210}_{82}Pb \xrightarrow{\beta} {}^{210}_{83}Bi + \gamma\text{–ray.}$$

Detection of radiations principle:

The radioactive radiation can be detected and measured by number of methods. The important one used in these days are given below.

Principle and working of different counters: Counters are the devices which are used for the detection and measurements of energy of radioactive radiations. All types of counters mainly based upon how different types of radiations interacts with the matter.

∝ - rays:

∝ - rays emitted from a particular nuclei will have almost same energies and they have different energies when they are emitted from different sources. In general, energy of ∝ - particles varies from 1meV to 10 meV. When ∝ - particles interacts with matter, they losses energy by two process namely.

1. By excitation process.
2. By ionization process.

5 meV of ∝ - particles produces about 20,000 ion pairs.

β - particles

β - particles also interacts with the matter and losses energy by excitation process and ionization process. β - particles causes less ionization of gaseous radicals when compared to ∝ - particle. Since they have small charge than ∝ - particle.

γ - rays

γ - rays causes primary ionization. The energy of γ - rays varies from 10keV to 7keV. They lose energy by three process namely (1) Photoelectric effect (2) Compton effect (3) pair production. In all these process, electrons are produced which in turn causes secondary ionization in the medium in which they are passed.

Detection of radiation :

Detection based on ionization :

When radiations falls on gaseous media, primary ionization takes place. The extent of primary ionization depends upon the energy of incident particles.

The different types of counters which works on the principles of ionization of gases are given below.

(a) Ionization chamber.

(b) Gieger muller counter.

When radiation passes through gaseous molecules taken in a chamber they cause, primary ionization as a result positive and negative ions are produced. When an external potential is applied, then +ve ions move towards cathode where as electrons get collected on anode wire and thus produces a current signal in the circuit. If the applied potential is increased, then ionization increases.

$$\left.\begin{matrix}\text{Current signal}\\ \text{produced}\end{matrix}\right] \propto \begin{matrix}\text{no. of ionpair}\\ \text{produced}\end{matrix} \propto \begin{matrix}\text{Energy lost by the particles}\\ \text{entering the system.}\end{matrix}$$

The number of electrons which are collected on anode wire varies with the applied potential as shown below. We get three characteristic regions.

When the potential is less than V_1, then the accelerating potential has no sufficient energy to separate ions and therefore recombination chances are more here.

$He^+ + e^- \gamma He.$

$He^+ + e^- \rightarrow He + \gamma.$

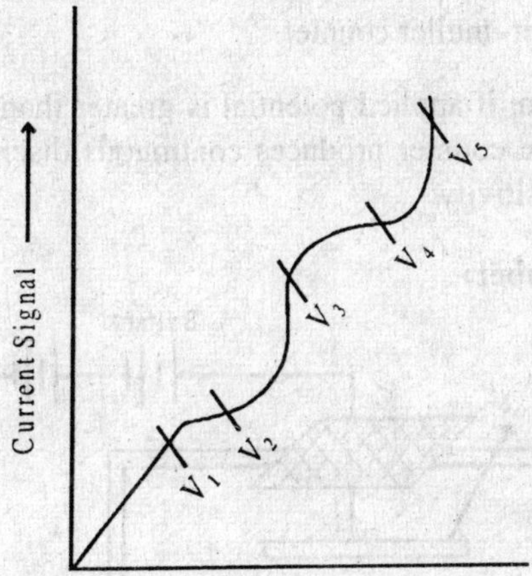

Fig. 5.4 : Variation of Current Signal in the detector with Voltage

In the region $V_1 - V_2$: The accelerating potential has sufficient energy to separate ions and so that, all the electrons get collected on anode wire. In this, recombination chances do not occur.

$$\left.\begin{array}{r}\text{Intensity of}\\ \text{Current pulse}\end{array}\right] \propto \begin{array}{c}\text{no. of ion}\\ \text{pairs produced}\end{array} \propto \begin{array}{c}\text{Energy lost by}\\ \text{entering particle}\end{array}$$

If current signal is weak, then it needs amplification. In this region, voltage fixing is not crucial and important. This is because, the small decrease or increase in voltage does not cause any change in number of electrons collected. The counter, which is operated in this region is called as "ionization chamber":

In the region $V_2 - V_3$: In this region, the electrons produced in primary process gets accelerated and causes secondary ionization at the applied potential. Therefore the current signal gets multiplied. In this region, no need to amplify current signal and hence the voltage fixing in this region is very important. The detector which is operated in this region is called as proportional counter, as current signal varies proportionally with the potential applied.

In the region $V_4 - V_5$: In this region, primary ionization, secondary ionization and tertiary ionization takes place and hence gives well defined current signal. In this region amplification of current signal is not required and counter which works in this region is called as Gieger–muller counter.

In this region, if applied potential is greater than V_5, discharge occurs and hence counter produces continuous discharge which is less than the sensitivity.

Ionization chamber:

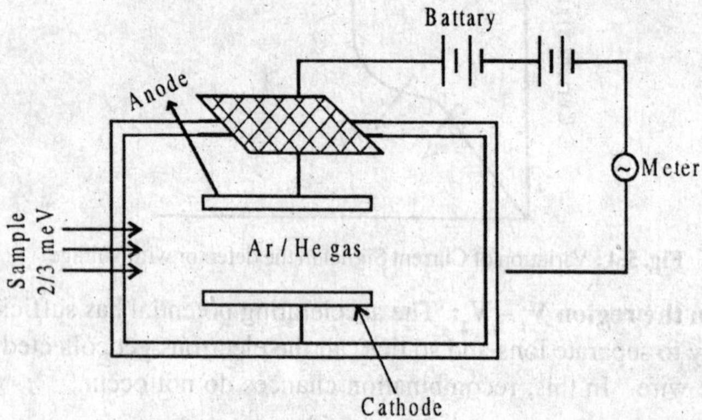

Ionization chamber works on the principles of ionization of gas molecules by the radiation entering into the system. The particles entering into the system causes primary ionization and external voltage is applied so that, ions get migrated towards different electrodes. If the applied potential is less than V_1 (minimum potential), then accelerating potential has no sufficient energy for separation of ions and hence, the chance of combinations of ions are more. On the other hand, applied potential is raised by a small amount ($V_1 - V_2$), then accelerating potential has sufficient energy to separate ions, so that different ions get collected on respective opposite electrodes and thus produces current signals.

The total charge (+ve or -ve) developed due to ionization is given

by $Q = \dfrac{Ze.ne \times 10^6}{30}$ eso/sec.

Ze is charge on each ion pair

n is number of particles entering the chamber

e is Energy lost in mev by moving particles.

30 ev is applied voltage required to produce anion pair.

Current pulse is directly proportional to number of primary ion pairs produced.

$$\begin{matrix}\text{Number of primary} \\ \text{ion pairs produced}\end{matrix} = \begin{bmatrix}\text{Specific ionization} \\ \text{energy}\end{bmatrix} = \begin{matrix}\text{dis tan ce between} \\ \text{two electrodes}\end{matrix}$$

Specific ionization energy is given by number of ion pairs produced in $10m^3$. Protons have high specific ionization and hence they require small ionization chamber, but β or γ - particles, have less specific ionization and so that, they need large ionization chamber.

There are two types of ionization chambers :

1. Integration type : In this type, there will be continuous accumulation of +ve charge.
2. Non – integration type : In this type each particle capable of causing ionization is recorded separately.

If the current signal is weak, the amplification is required. Meter in the ionization chamber is not enough sensitive to measure the energy of individual particles entering into the system, though it can integrate number of particles per second. So in order to measure the energy of individual particles, this arrangement is used.

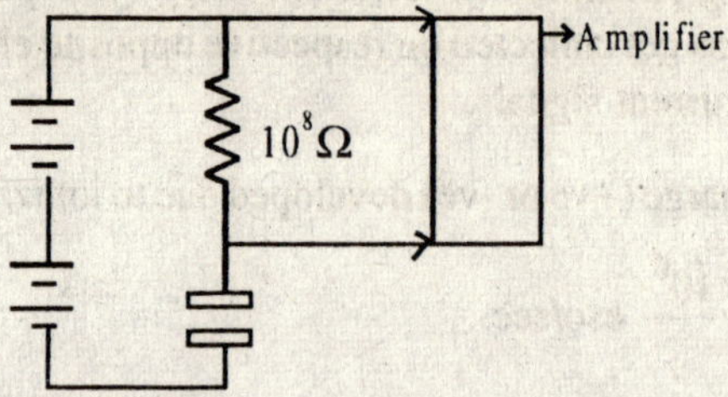

Due to high resistance in the circuit. Voltage drop gives well defined current signal. Here, even a small current produced by a single particle can be measured.

Gieger – muller counter or G.M. – counter

Gieger – muller counter can't be used as detector for the detection of ∝ and β - particles entering into the system but can be used for the measure of energy of radiations entering into chamber. It also works on the principle of ionization of gas by incoming particles.

It works in the high voltage region (ν_4 - ν_5) and hence no need of amplification of current signal. In this region primary, secondary ionization and tertiary ionization takes place.

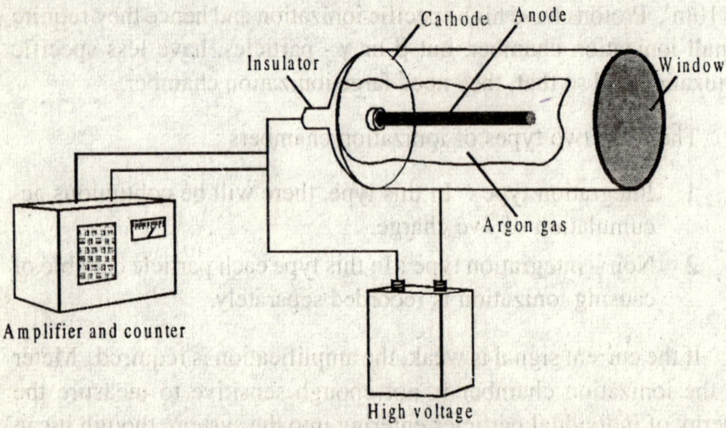

Fig. 5.4 : Gieger Muller Counter

The radiation particle entering into counter, causes ionization of gas ($Ar^+ + e^-$) as $m^+ + e^-$ and these electrons get accelerated by the applied potential and these accelerated electrons cause secondary ionization. Due to this ionization, avalanche of ionization or collision ionization takes place.

Suppose if secondary process is allowed to continue, secondary discharge occurs at cathode. So that, no current signal was obtained. To avoid this quench, and the process of avoiding the contribution of secondary process at some stage by using a particular quenching agent is called quenching process.

The most commonly used quenching agent is 10% alcohol.

$$M^+ + \text{alcohol} \rightarrow M_{\text{metal}} + \text{alcohol}^+.$$

Alcohol$^+$ again dissociates into uncharged fragment. If it is not quenched at some stage, then continuous discharge takes place and no current signal.

Since the positive charges moves with a low velocity compare to electron, a thin sheath of wire for about 10^{-3} to 10^{-4} second which results in the decrease of applied potential. So that, secondary particles entering into counter chamber cannot be detected and that time interval is called as 'dead time' of counter.

The mechanism which occurs in G.M. counter is given below.

1. Production of ion pair $M \rightarrow M^+ + 2e^-$.
2. Transport of ions near anode line [secondary process] $M^+ + e^- \rightarrow M^+ + 2e^-$.
3. Partial recombination of electron and ions. $M^+ + e^- \rightarrow M + h\nu$.
4. Production of new ion pair or photoelectron. $M + h\nu \rightarrow M^+ + e^-$.

G.M. counter is more useful because,

1. It is highly sensitive.
2. Gives well defined signals
3. No amplification is needed.

Scintillation counter:

One of the earliest detector used for the measurement of energy of radiations is called as scintillation counter.

Scintillation counters can be used for the measurement of energy of ∝ - and γ - ray particles but not for β - particles.

Principle : When radiations are made to fall on scintillating material, it emits flash of light and it is made to fall on photo cathode surface from which number of photoelectron are emitted and the intensity of emitted photo electrons depends upon energy lost by the radiation. The scintillation material used may be solid or liquid. Example : anthracene, terphenyl, stillbene etc.

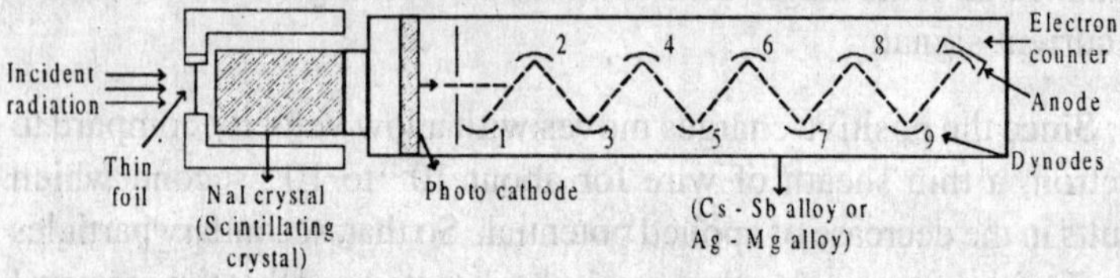

Fig. 5.5 : Scintillation Counter

Sodium Iodide (No. 1) is stored in air tight bottles because it is highly hygroscopic. NaI is mixed with the 0.5% TI which acts as activator. It is having a size of 10 inches diameter and 10 inches length.

The radiations are made to fall on scintillating material, NaI which is activated with small amount of TI. The flashes of light (3000 – 500 hν) emitted by scintillator and are made to fall on photo cathode surface, so that photo electrons are ejected and are accelerated towards dynodes, which are coated with Ag – Mg alloy. In this process electrons of different energies are produced for each incident scintillation and again they are accelerated towards dynode – 2 and similar process is repeated towards d_3, d_4 —— d_9 and finally electrons gets collected on anode which gives amplified current signal. Current signal produced is proportional to energy last by incident particle. If the number of electrons emitted at each dynode is 'n' times greater than that of incident electron, then for 9th dynode, amplification factor is n^9. Using this counter, radiation of upto 10^3 disintegration per second can be found out accurately.

Radioactive isotopes and their use as tracers:

Isotopes are the different atoms of same element having same atomic number but different atomic weights. Isotopes of the same element which are radioactive are called as radioactive isotopes.

Eg.: i) $^{238}_{92}U$, ii) $^{235}_{97}U$ and

i) $^{234}_{90}Th$, ii) $^{232}_{90}Th$,

Isotopes of same element which are non – radioactive are called as non – radioactive isotopes.

Eg.: i) $^{2}_{1}H$, ii) $^{1}_{1}H$ and

i) $^{206}_{82}Pb$, ii) $^{207}_{82}Pb$, iii) $^{208}_{82}Pb$.

Since, the isotope possess the same atomic number, they contain same number of protons and electrons but they contain different number of neutrons in their nucleii. Therefore, the isotopes possess same chemical properties but different physical properties.

Radioactive isotopes as tracers :

Radioactive isotopes are frequently used as tracers or tagged atoms to study many processes.

When we are using radio isotopes as tracers following points to be considered.

1. The half – life of isotopes used for tracer technique should not be neither short nor long. The half life should be in between months and year.
2. For the detection of radiation only GM counter or scintillation counters are used.

The use of radio isotope as a tracer in tracer technique is based on the fact that, the isotopic species of elements are chemically similar. Hence they are distinguished by their e/m ratio, and because the one introduced is radioactive it can be determine by its radioactivity.

An isotope of element used for this purpose is called as tracer and the element which is labelled with a particular isotope is called as tracer compound. The basic principle involved in tracer technique is to mix the material under investigation in a chemical reaction with a suitable tracer and after certain interval of time, small portion of reaction mixture is taken out and then analyzed.

At present 274 stable nucleus and 2000 radioactive isotopes are known. Out of 2000, only 100 – 150 radioactive isotopes find wide applications reported for the investigation of problems in various areas of science such as medicine, agriculture and industries etc. Tracer studies have provided a very valuable information about the movement of atoms, ions and molecules in gaseous liquid and solid phases. They are used in number of reactions to study mechanisms of both organic and inorganic chemistry rearrangement reactions, polymerization reactions etc.,

Some of radioactive isotopes are given below.

^{14}C, ^{128}I, ^{141}Zr, ^{38}Cl, ^{59}Fe, ^{32}P, ^{60}Co, ^{165}Zn, ^{24}Na, ^{51}Cr ^{3}H.

Commonly used radioactive isotopes as tracers are ^{14}C, ^{32}P, ^{24}Na, ^{38}Cl, ^{131}I, ^{60}Co.

Sources of radioactive isotopes

1. Natural sources: When neutrons from cosmic radiations bombarded with atmosphere nitrogen gives ^{14}C isotopes.

$$3\ {}^{1}_{0}n + {}^{14}_{7}N \rightarrow {}^{14}_{6}C + {}^{3}_{1}H$$

2. Nuclear reactions: Fission products of nuclear reactors.

   When a heavy nucleus is broken by nuclear reactions, smaller radioactive elements are formed

Eg.: $${}^{235}_{92}U + {}^{1}_{0}n \longrightarrow {}^{139}_{56}Ba + {}^{94}_{36}Kr + 3\,{}^{1}_{0}n + \text{Energy}$$

1. <u>Decay of natural radioactive series:</u> In the naturally occurring radioactive series, a number of radioactive isotopes are formed before getting a stable isotope of lead.
2. <u>Nuclear reactions:</u> In some of the nuclear reactions, radioactive isotopes are produced can be used for tracer studies.

$$^{31}_{15}P + ^{2}_{1}H \rightarrow ^{32}_{15}P + ^{1}_{1}H$$

$$^{27}_{13}Al + ^{1}_{0}n \rightarrow ^{24}_{11}Na + ^{4}_{2}He$$

Preparation of radioactive isotopes:

1. ^{14}C isotope : Half life of ^{14}C isotope is 5730 year. It is obtained in the nature by simple reaction of cosmic ray neutrons with atmospheric nitrogen.

$$^{1}_{0}n + ^{14}_{7}N \rightarrow ^{14}_{6}C + ^{1}_{1}H$$

^{14}C is a β – emitter.

$^{14}_{6}C$ an crotope.

Laboratory Method:

The radio labeled carbon was obtained by bombarding the neutrons in a nuclear reactor with target Be_3N_2 which is dissolved in 65%. H_2SO_4 and H_2O_2. On bombarding slow flux of neutrons, $^{14}CO_2$ is formed along with a small amount of ^{14}CO, H ^{14}CN and $^{14}CH_4$, on passing this mixture of gases over heated CuO at 750°c, $^{14}CO_2$ get adsorbed on the surface of CuO. Then it is treated with 60% NaOH to get $Na_2{}^{14}CO_3$ and then when this product is treated with $Ba(OH)_2$, $Ba^{14}CO_3$ Precipitated out.

2) <u>^{24}Na – isotope:</u> It is also a β - emitter. The half life of ^{24}Na is 15.03 years. It is generally obtained by the bombardment of $^{23}Na_2CO_3$ with slow neutrons in nuclear reactor.

$$^{23}Na_2\,CO_3 + ^{1}_{0}n \rightarrow ^{24}Na_2CO_3 + \gamma$$

3) <u>^{32}P – isotope:</u> It is also a β emitter and has a half life 14.23 days. It is produced in nuclear reactor by neutron reaction.

$^{31}_{15}P + ^{1}_{0}n \rightarrow ^{32}_{15}P + \gamma$ particle

It is also be obtained by laboratory method. When $^{32}_{16}S$ is bombarded with slow moving neutrons, $^{32}_{15}P$ is formed. It is then dissolved in dil. HNO_3, to oxidize other impurities and it is treated with aquaregia, and finally with lanthanides, to get $H_3{}^{32}PO_4$ which is precipitated out as lanthanum salts. The precipitate dissolved in HCl and passed through cation exchanger to remove lanthanum $^{32}PO_4^{3-}$ is obtained in solution.

Applications of radio active isotopes:

Both radioactive and non-radioactive isotopes play an important role in different fields. The radioactive isotopes can be used directly or indirectly by introducing these into system under investigation.

1. In a chemical Investigation: Radioactive isotopes finds a wide application in chemistry.

a. To study reaction mechanism: For examples oxidation of CO to CO_2. In this reaction, MnO_2 is only used as catalyst. Earlier, it was believed that, CO, using oxygen of MnO_2 oxidized to CO_2.

   But $CO + (O)_{air} + MnO_2 \rightarrow CO_2 + [MnO_2]$

   This reaction indicates that, when CO is made to react with air in presence of labelled, MnO_2. All the activity remains with MnO_2 and, CO_2 produced is inactive. It indicates that, oxygen atom of MnO_2 is not involved in the oxidation of CO to CO_2, but it is that of atmosphere Oxygen.

   Esterification reaction:

   $R\,CH_2\,COOH + HOCH_3 \rightarrow H_2O + R\,CH_2\,COO\,CH_3$

   To know whether the water formed during esterification reaction is from H of acid and –OH of alcohol or OH of acid and H of alcohol, the following reaction was carried out. The radio labeled oxygen in alcohol was used. The ester formed in the reaction and water was analyzed for radioactivity.

Water molecule formed shows no activity which indicates that, water molecule is formed from –OH of acid and –H of alcohol.

2. Structural investigation of chemical compounds

   Structure of PCl_5. trigonal bipyramidal structure was proposed for PCl_5.

$$PCl_3 + Cl_2 \rightarrow PCl_5 \xrightarrow{\text{Hydrolysis}} POCl_3 + HCl$$

   The Cl –atom in HCl differ from those present in $POCl_3$. All the activity remains only with HCl. Thus trigonal bipyramidal structure of PCl_5 was confirmed by tracer technique applications.

3. Medical applications of radioactive isotopes

a. Detection of blood circulation.

   To check the restriction for the supply of blood to different parts of body, ^{24}Na was taken in the form of saline solution and injected into the body. It requires definite time for the circulation of this solution to different organs of body. If it requires more time than prescribed time, it shows maximum activity, which indicates restriction for the flow of blood in the particular part of body.

b. In the treatment of cancer. ^{60}Co can be used for treatment of cancer disease. Since this is a β - emitter, it specifically destroy the cancer cells.

4. Agricultural applications. Radioactive isotopes also finds a large number of application in agricultural field. They can be used to check the stability of fertilizer depending upon nature of plant and soil. For acidic soil the $*CaSO_4$ and $*CaCO_3$ is used along with regular fertilizer. The fertilizer is left over in the soil for two days and check for the uptake of *Ca by measuring its activity in the body of plants. The uptake *Ca depends on acidic or basic nature of soil. Similarly the uptake of phosphate and nitrogenous fertilizers can be checked for their stability.

5. Industrial application: In industries, they can be used as catalyst and also to check the wear and tare of machine parts.

Here, we activate the moving parts of machine tools, such as piston, tools, etc., by bombarding with neutrons. After a few days, collect the known amount of grease and analyze the sample for its activity.

By knowing the activity shown by the grease, it is possible to know how much of wear and tare taking place. More the activity of grease, more will be its wear out.

Age of minerals and rocks: we have seen that in any radioactive disintegration series, the final product formed is isotopes of lead. Thus by determining the amount of parent radioactive element and the isotope of lead in a sample the age of rock can be calculated. We assume that the rock did not contain any lead isotope initially.

Problem: A samples of pitch blende contains 0.05 g of Pb^{206} for every gram of U^{238} present in it. Calculate the age of the rock ($t_{1/2}$ of $U^{238} = 4.5 \times 10^9$ years)

Solution: $t_{1/2} = 4.5 \times 10^9$ years

$$\lambda = \frac{0.693}{4.5 \times 10^9} \text{ year}^{-1}$$

Amount of U^{238} present at the time of analysis (N) = 1 g

Amount of U^{238} present originally (N_0)

$$= \left\{\begin{matrix}\text{Amount of } U^{238} \\ \text{now present}\end{matrix}\right\} + \left\{\begin{matrix}\text{Amount of } U^{238} \text{ changed} \\ \text{to } Pb^{206}\end{matrix}\right\}$$

$$= 1 + \frac{238}{206} \times 0.05\text{g}$$

$$= 1.0577\text{g}.$$

Let 't' be the time in which 1.0577 g of U^{238} is reduced to 1g of U^{238}.

$$t = \frac{2.303}{\lambda} \log \frac{N_0}{N_t}$$

$$t = \frac{2.303}{0.693} \times 4.5 \times 10^9 \times \log \frac{1.0577}{1} = 3.46 \times 10^8 \text{ years}$$

Carbon – 14 dating: By this the age of archeological objects (wood, dead plants, animals) can be determined. This was developed by Willard Libby.

The principle is the cosmic rays bombard with nitrogen atoms present in atmosphere forming radioactive C – 14.

$$^{14}_{7}N + ^{1}_{0}n \longrightarrow ^{14}_{6}C + ^{1}_{1}H$$

This is oxidized to CO_2. Thus $^{14}CO_2$ present in atmosphere is assimilated by plants and animals. Living plants and animals have a definite and constant proportion of ^{12}C and ^{14}C. When plant or animal dies no fresh ^{14}C enters the body, it starts decaying as $^{14}_{6}C$ = 5770 years.

The amount of ^{14}C in a sample can be accurately determined by counting the number of β - particles emitting per minute by one g of the sample through this the age of sample can be found out.

Problem: A sample of wood from archeological find has 5 counts |minute| gm carbon. Calculate the age of wood sample. A freshly cut wood gives 15.3 counts |minute| gm of carbon. Half – life of ^{14}C - 5770 years.

Solution: $t_{1/2} = \frac{0.693}{\lambda}$ or $\lambda = \frac{0.693}{t_{1/2}} = \frac{0.693}{5770} = 1.202 \times 10^{-4} \text{ yr}^{-1}$.

Let 't' be the age of the sample

$$t = \frac{2.303}{\lambda} \log \frac{N_0}{N_t} = \frac{2.303}{1.201 \times 10^{-4}} \log \frac{15 - 3}{5} = 9313 \text{ years.}$$

Nuclear fission: the process of splitting of a heavier nucleus into fragments by bombarding with suitable projectiles is called 'Nuclear fission'.

Amount of energy is released during fission due to disappearance certain mass according to the equation $E = mc^2$.

Energy released = mass defect x 931.48mev. For every uranium atom of fission 261.75 mev or 8 x 10^7 kJ energy is released.

Concept of critical mass: the minimum amount of fissionable material required so as to continue chain reaction is called critical mass. If the mass is more than critical mass it is called super critical mass and if the mass is less than the critical mass it is called sub-critical mass. For uranium the critical mass is between 1 kg and 100 kg.

Nuclear Reactors: the device in which the nuclear reaction is carried out in a controlled manner is called nuclear reactors. The main parts of nuclear reactors are as shown below.

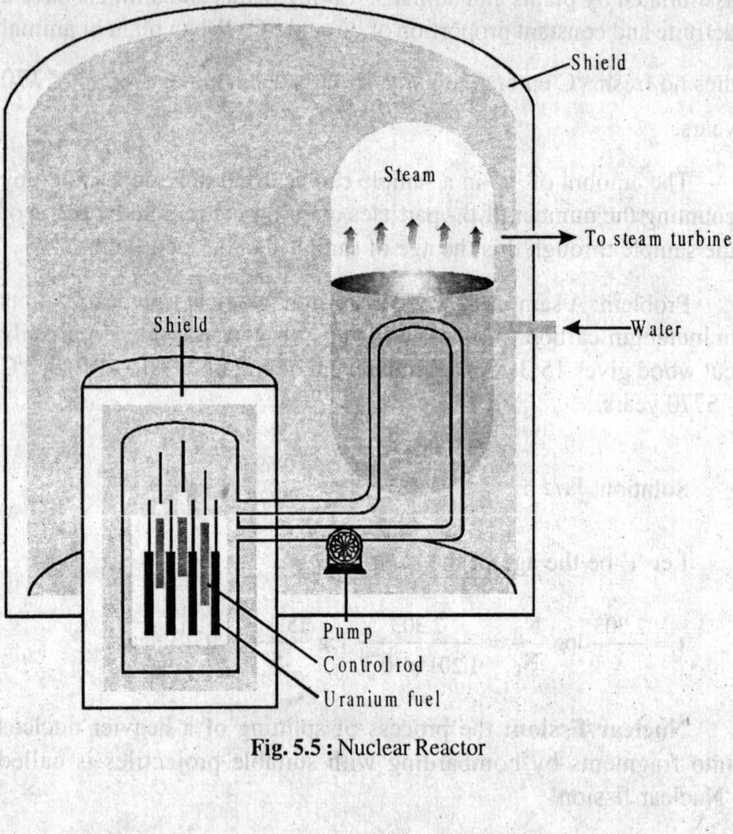

Fig. 5.5 : Nuclear Reactor

1. **Fuel rods:** the fissionable material used in the reactor is called fuel. The fuel used is enriched U^{235}, in the form of rods or pellets, shielded by steel tubes.
2. **Control rods:** To control the fission process, rods made of cadmium or boron are introduced along with the fuel rods. These rods can be raised or lowered to control the fission process by absorbing neutrons.
3. **Moderator:** the fission process is most efficient with slow neutrons, heavy water will function as moderator.
4. **Coolant:** the heat liberated during the fission process will be absorbed or taken up by coolant is used to convert water into steam this steam rotate the turbine to produce electricity. Sodium is used as coolant.
5. **Shield:** to avoid the loss of heat and for the safety of the person working, the reactor core is shielded with steel or concrete dome called the shield.
6. **Breeder Reactors:** the non fissionable materials like U^{238} is converted into fissionable materials like plutonium – 239 and uranium – 233 with neutrons. The reactors in which this conversion is carried out are referred to as Breeder reactors

$$^{238}_{92}U + ^{1}_{0}n \longrightarrow ^{239}_{92}U \longrightarrow ^{239}_{93}Np + ^{0}_{-1}e$$

$$^{239}_{93}Np \longrightarrow ^{239}_{94}Pu + ^{0}_{-1}e$$

$$^{239}_{94}Pu + ^{0}_{-1}n \longrightarrow \text{fission products + neutrons}$$

The non – fissionable nuclide like U^{238} which can be converted to fissile nuclide is called fertile nuclide.

Nuclear fusion: the process in which lighter nuclei fuse together to form a heavier nucleus is called nuclear fusion.

$$^{2}_{1}H + ^{2}_{1}H \longrightarrow ^{4}_{2}He \quad 0.026 \text{ mass decrease} \quad (23 \times 10^{23} \text{ kJ / mole})$$

$$4^{1}_{1}H \;\; ^{4}_{2}He + 2\,^{0}_{+1}e \quad 0.029 \text{ mass decrease} \quad (26 \times 10^{8} \text{ kJ / mole})$$

The nuclear fussion reactions require very high temperature $>10^6$K, in order to overcome electrostatic repulsion between nuclei when they come together to fuse. They are called thermonuclear reactions. The energy released / gm of the mass in nuclear fusion is greater than that in fission process.

Difference between fission and fusion

Nuclear fission	*Nuclear fusion*
1. Involves splitting of heavy nucleus into smaller fragments.	1. Involves fusion of lighter nuclei into heavier nuclei.
2. Chain process	2. Not a chain process
3. High temperature is not essential	3. High temperature is essential
4. Can be controlled	4. Difficult to control
5. Nuclear waste which is radioactive.	5. No nuclear waste

QUESTIONS

1) Define radioactivity give a example of radioactive element.
2) What is Natural radioactivity? Give one example.
3) What is Artificial or induced radioactivity? Give an example.
4) How is the stability of the nucleus depends on ratio.
5) What is group displacement law of radioactivity.
6) Describe the nature and properties of the rays emitted by radioactive materials.
7) Calculate the number of α^- and β^- particles emitted when changes into.
8) Explain the theory of radioactive disintegration.
9) Write a short notes on isotopes.
10) Write a note on mass defect.
11) Write a note on nuclear fission.
12) Give a method of preparation of radio isotope.
13) Distinguish between natural and in dived radioactivity.
14) What is nuclear fusion?
15) Describe the important any two applications of radioactivity.

16) What is meant by tracer and tracer technique?
17) Discuss any two applications of tracer and tracer technique.
18) What is a nuclear reactor? Give the design of the nuclear reactor.
19) Describe a method for the determination of radioactivity.
20) What is a binding energy? How it is calculated?
21) What is a mass defect? How it is calculated?
22) How do you explain the stability of atom with reference to ratio?
23) What are isotopes? Give an example.

6

SOLUTIONS AND COLLIGATIVE PROPERTIES

SOLUTION: A solution may be defined as a condensed fluid and homogenous system of not less than two components, solute and a solvent.

SOLUTE: The substance which dissolves in a solvent is referred to as solute.

SOLVENT: The substance in which the dissolution of the solute takes place is referred to as solvent. In a given solution a larger portion of it is the solvent and the lesser portion is the solute.

SATURATED SOLUTION: At a given temperature a solution contains as much solute as it can hold in the presence of the dissolving substance is said to be saturated solution.

UNSATURATED SOLUTION: At a given temperature any solution which contains less than the required amount of the substance (or solute) is unsaturated. If it is present more than this amount it is called supersaturated.

Although solutions with many components can be prepared, the solutions of two components are of great interest as it becomes easy to understand and also deal with them. The various types of solutions are:

A. Solution of gas in a gas. Example: Air, cooling gas.

B. Solution of a liquid in gas. Example: Moisture in air

C. Solutions of a solid in gas. Example: Camphor in air

D. Solution of a gas in solid. Example: Hydrogen in palladium. Adsorption of gas on solid.

E. Solution of a liquid in solid. Example: Jelles

F. Solution of a solid in a solid. Example: Alloys, brass, gold ornaments containing copper etc.

G. A solution of a gas in a liquid. Example: Chlorine Gas in water.

Example: 1) CO_2 gas in aerated water

2) NH_3 gas in aerated water

3) H_2S gas in aerated water

H. Solution of solid in a liquid. Example: 1) Sugar in H_2O

2) Table salt in H_2O

I. Solution of a liquid in a liquid Example: C_2H_5OH in H_2O

1. Units of Concentrations

The concentration of the solution is the quantity of the substance (solute or solid) dissolved in a known amount of the solvent. The concentrations can be expressed in different ways:

1.1 Percentage

The amount of a solution is usually expressed either as percentage by weight or percentage by volume. If nothing is mentioned then it is percentage by weight.

a) Liquid Dissolved in Another Liquid

Percentage by weight has the same meaning, but percentage by volume means the volume of the liquid solute in cm^3 present in 100 cm^3 of the solution. For instance 5% solution of ethanol in water by weight means 5g of ethanol dissolved in 100g, of solution [i.e. 5g ethanol + 95 cm^3 H_2O = 100 cm^3 solution] Thus, 5% ethanol solution means that 5 cm^3 by volume of ethanol are dissolved in 100 cm^3 of the solution.

b) Solid Dissolved in a Liquid

In this case, the percentage by weight means that, the solid (solute) in grams present in 100 gram of the solution and percentage by volume means that the weight of the solute dissolved in 100 cm^3 of the solution. For example, 10% sugar solution by weight means that 10g of sugar present in 100 gram of its solution. Similarly 10% solution of sugar by volume means that 10 gram sugar dissolved in exactly 100 cm^3 of the solution.

1.2 Strength of the Solution

It is defined as the amount of the solute in grams dissolved in one liter (or 1 dm^3) of the solution and hence it is expressed in g/liter(or g/dm^3)

1.3 Molarity of the Solution

It is defined as the number of moles of the solute dissolved in 1litre (or a dm^3) of solution. It is denoted by 'M'

Mathematically, it is expressed as

$$M = \frac{\text{Number of Moles of solute}}{\text{Volume of the solution in litre (or dm}^3)}$$

The value of M can be calculated as follows:

Let us consider X grams of a solute present in V cm^3 of a given solution, then

$$\frac{M}{\text{Molar Mass}} \times \frac{1000}{V}$$

For example, a solution of H_2SO_4 acid having 4.9 grams of it dissolved in 500 cm^3 of its solution should have its molarity given by

$$\text{Molarity} = \frac{\text{Wt. of the solute per liter}}{\text{Molecular Wt.}}$$

$$M = \frac{4.9}{98} \times \frac{1000}{500} = 0.1\ M$$

1.4 Formality of a Solution

This concept is used in the case of ionic compounds such as NaCl, KCl $(NH_4)_2SO_4$, Na_2CO_3. It is used in place of molarity.

It is defined as the number of gram formula mass of the solute dissolved in 1 liter of the solution, formula weight of the substance is used to express its concentration unit as F. the term formula mass or weight is used in place of molecular mass (or weight) because ionic compounds exist as ions and not as molecules.

For example, it is important to note that an aqueous solution of 0.2M Na_2CO_3 contains 0.4M Na^+ ions and 0.2 M CO_3^{2-}ions. (The aqueous solutions containing 1 mole of Na_2CO_3 gives 2 moles of Na^+ ions and 1 mole of CO_3^{2-} ions).

1.5 Normality of a Solution

It is defined as the number of gram equivalents of the solute dissolved in 1 litre (or 1 dm^3) of given solution. It is represented by a symbol 'N'

Mathematically, it is expressed as

$$N = \frac{\text{Number of gram equivalent of solute}}{\text{Volume of the solution in litre (or dm}^3\text{)}}$$

The value of N can be calculated from the strength of a solution as follows:

Let us consider 'X' grams of a solute present in V cm^3 of a given solution, then

$$N = \frac{X}{\text{eq. weight of the solute}} \times \frac{1000}{V}$$

For example, a solution of H_2SO_4 having 0.49 gram of it dissolved in 250 cm^3 of its solution should have a normality given by

$$N = \frac{0.49}{49} \text{ x } \frac{1000}{250} = 0.04$$

(Note: Equivalent weight of H_2SO_4 = 49)

1.6 Molality of a Solution

It is defined as the number of moles of the solute dissolved in 1000 grams of the solvent. It is denoted by m.

Mathematically, it is expressed as

$$m = \frac{\text{Number of moles of solute}}{\text{Weight of the solvent in grams}} \times 1000$$

The value of m can be calculated from the strength of a solutions as follows:

$$m = \frac{\text{Strength of a solution}}{\text{Molecular weight of the solute.}}$$

Let us assume that x grams of the solute be present in y grams of the solvent. then,

$$m = \frac{X}{\text{Molecular weight of the solute}} \times \frac{1000}{y}$$

For example, a solution of anhydrous Na_2CO_3 (Molecular weight = 106) having 1.325 grams of it dissolved in 250 grams of water should have its molality given by

$$m = \frac{1.325}{106} \times \frac{1000}{250} = 0.05\,m$$

1.7 Mole Fraction of a Solute

Mole fraction of a constituent (solute as well as solvent) is the fraction obtained by dividing the number of moles of that constituent by the total number of moles of all the constituents present in the solution. It is usually denoted by 'x'.

In a given solution, if n_1 and n_2 are the number of moles of the solvent and solute, respectively, then the mole fractions of the solvent (x_1) in the solution can be calculated as $x_1 = \frac{n_1}{n_1 + n_2}$ and the mole fraction of the solute (x_2) in the solution can be calculated as

$$x_2 = \frac{n_2}{n_1 + n_2}$$

For example, if a solution contains 4 moles of C_2H_5OH and 6 moles of H_2O, then the mole fraction of the solute

$$C_2H_5OH = \frac{4}{4+6} = \frac{4}{10} = 0.4$$

and mole fraction of solvent water $= \frac{6}{4+6} = \frac{6}{10} = 0.6$

In general, if a solution contains a number of components (say i) fraction of different components will be

$$x_1 = \frac{n_1}{n_1 + n_2 + n_3 \ldots\ldots + ni} = \frac{n_1}{\Sigma ni}, \text{ and}$$

i = 1

$$x_2 = \frac{n_2}{n_1 + n_2 + n_3 \ldots\ldots + ni} = \frac{n_2}{\Sigma ni}$$

i = 1

and so on.

The sum of the mole fraction of all the components (constituents) present in a given solution should be equal to 1. It may also be noted that the mole fraction is a dimension quantity.

1.8 Mass Fraction of a Component

It is defined as the mass of the given component per unit of the solution.

Let us consider that x_A and x_B be the mass fractions of the two components A and B, respectively. Let W_A grams of the component A be mixed with W_B grams of the other component B in a binary solution. Then,

$$x_A = \frac{W_A}{W_A + W_B} \text{ and } x_B = \frac{W_B}{W_A + W_B}$$

Therefore, evidently it follows that $x_A + x_B = 1$

mass fraction multiplied by 100 gives mass percentage. In case of component A, the mass percentage is given by

$A = x_A \times 100$, and similarly

$B = x_B \times 100$

It may be noted that like mole fraction, mass fraction is also a dimension less quantity.

1.9 Concentration Expressed in Parts Per Million (PPM)

The term ppm is used to express the concentration of a solution. It denotes that a small amount of the solute in a very large amount of the solution (or solvent)

It is defined as the mass of the solute dissolved in one million (1×10^6) parts by mass of the solution.

For a given solute A, its concentration is expressed as

$$\text{ppm} = \frac{\text{Mass or (weight) of A}}{\text{Mass (or weight) of solution}} \times 10^6$$

Generally, the pollution of the air (atmosphere) is reported in ppm, but it is expressed in terms of volume rather than mass, i.e., the volume of the harmful gas (e.g., SO_2) in cm^3 Present in 10^6 cm^3 of the air.

2.0 Some Important Relationships Between the Various Concentration Terms

2.1 Relationship Between Volume and Normality or Molarity of a Solution.

The equation used in this case is given by $N_1 x V_1 = N_2 x V_2$

Where N_1 is the normality of the solution-1 and V_1 its volume. When either N_1 or V_1 is changed, new normality N_2 and its volume

V_2 are calculated, then the above relationship is called normality equation. It is also applicable when V_1 cm^3 of another solution and normality N_2.

In the same manner, if molarity and volume of a solution are changed from $M_1 \times V_1$ to $M_2 \times V_2$ then

$$M_1 \times V_1 = M_2 \times V_2 \text{, is called molarity equation.}$$

It is good to note that in a balanced chemical equation, if n_1 moles of reactant react with n_2 moles of reactant 2, then

$$\frac{M_1 \times V_1}{n_1} = \frac{M_2 \times V_2}{n_2}$$

2.2 When Volumes of Certain Concentration Mixed

If a volume V_1 cm^3 of a solution of normality N_1 is mixed with the volume V_2 cm^3 of normality N_2 of another solution. Then the resultant normality N_3 of the final solution is given by

$$N_1V_1 + N_2V_2 = N_3(V_1+V_2)$$

$$N_3 = \frac{N_1 V_1 + N_2 V_2}{(V_1 + V_2)}$$

$$\text{Similarly,} \quad M_3 = \frac{M_1 V_1 + M_2 V_2}{(V_1 + V_2)}$$

2.3. Relationship Between Molarity (M) and Molality (m)

The relationship between Molarity and Molality of a solution is given by an equation.

$$m = \frac{M \times d}{(1 + MM_2 / 1000)}$$

Where M is the molarity,

d is the density of the solution

M_2 is the molar mass of the solute.

m – molality

2.4. Relationship Between Molality (M) and Mole Fraction (X_2)

The relationship between Molality and Mole fraction is given by the equation.

$$m = \frac{1000\ x_2}{x_1 M_1}$$

Where m is the molality of the solution.

M_1 is the molar mass of the solvent, and

X_1 is the mole fraction of the solvent.

2.5 Relationship Between Molarity (M) And Mole Fraction (X_2)

The relationship between Molarity and Mole fraction is given by an equation.

$$M = \frac{1000\ d\ X_2}{X_1 M_1 + X_2 M_2}$$

Where M is the Molarity of the solution

d is the density of the solution.

X_1 is the mole fraction of the solvent.

X_2 is the mole fraction of the solute.

M_1 is the molar mass of the solvent.

and M_2 is the molar mass of the solute.

2.6 Dilution of a Solution

This is an important step in many chemistry experiments where a given solution need to be diluted.

If $V_1 cm^3$ of a solution with molarity M_1 need to be diluted to have molarity M_2 so that the final volume is V_2, then, volume of water/solvent to be added will be $(V_2 - V_1)$ and is given by

$$(V_2 - V_1) = \left(\frac{M_1 - M_2}{M_2}\right)^{V_1}.$$

when a solute is added to a solvent, two things can happen to the solute.

i. It can move towards the increased randomness or entropy.

ii. It can move towards decrease in energy.

These two factors are opposite to each other.

Randomness tries to increase the solubility while energy factor tries to decrease it. If the former factor prevails, the substance dissolves, if the latter prevails, it does not become soluble.

For example, When NH_4NO_3 is dissolved in H_2O the solution is found to be cooler than water taken in container, then NH_4NO_3 have gained energy from water during dilution. This tendency is overcome by randomness produced due to ionisation of this compound. The greater randomness prevails on the energy factor and the compound dissolves.

How a Solid Substance Dissolve in a Solvent

The solids can be classified into ionic or non-ionic solids. Those contains positively charged and negatively charged ions are called non-ionic solids. It is the force of attraction between oppositely charged ions in a crystal which gives the lattice energy of the crystal. It is the lattice energy which opposes the tendency of a solute to dissolve. The process of dissolution of a solid substance in a solvent is explained by the electrical forces operating between the molecules or ions of solvent.

Eg: when NaCl dissolves in H_2O.

Solubilities of Non -ionic Substances

Many non - ionic substances dissolve in polar solvents like H_2O.

Eg: Sugar dissolves in H_2O, which consist of a number of polar OH^- groups. The hydrogen$_2$ bonding takes place through hydrogen groups of the sugar moleules to dissolve it in the solvent

Factors Affecting the Solubility

The solubility of a solid in a given solvent is influenced by mass factors. These factors are -

a) Temperature
b) Nature of solute
c) Nature of solvent.

a) Temperature

It has been observed that the solubility is a function of the temperature factor. The temperature increases the solubility of the solute in a given solvent. The solubility curves for some inorganic compounds in H_2O, at different temperature are given below.

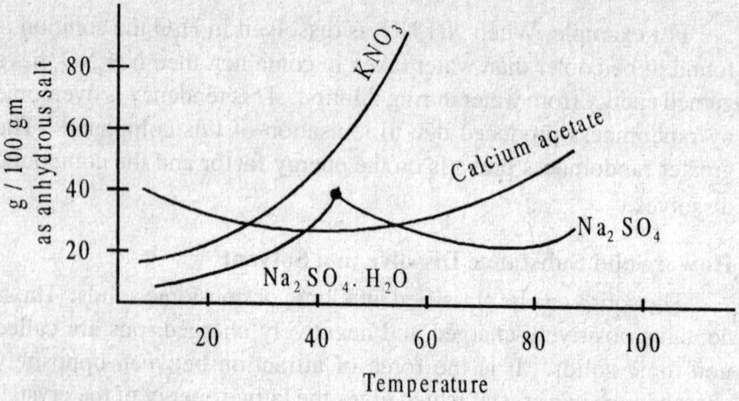

The solubility of solute like calcium acetate decreases with rise of temperature while in the case of solute KNO_3, the solubility increases. This behaviour in the solubility curves can be explained on the basis of **Le - Chatelier's principle** when, "a mole of the substance is dissolved completely in a certain excess volume of solvent, the change in heat content is called the enthalpy of solution".

If the enthalpy of a solution is +ve, then it favours the solubility of the solute. If the rise in temperature of the solution, which decreases the solubility. In the case of Na_2SO_4, $10H_2O$. The solubility curve is found to be have a break indicating a transition of the solid.

b) Nature of Solute

The nature of the solute plays a greater role in the solubility behaviour. When ionic solids are dissolved, ionizes into ions. The ions are held together by a force of attraction (F) between the oppositely charged ions is give by ; $F = \frac{f^+ + f^-}{r^2 D}$.

Where f + and f- are the charges, r is the distance between the charge and D is the dielectric constant of the medium. If the force of attraction increases, the product of the charges increases, and therefore solubility decreases. The product of charges in ($Na_2 SO_4$) bi – univalent or bi – bivalent ($Mg SO_4$, $Zn SO_4$) or $(NH_4)_2 SO_4$ are higher than the uni – univalent electrolytes (like NaCl, KCl, RbCl etc.). Hence, bi – univalent electrolytes (like $BaCl_2$), bi – bivalent electrolytes like ($Ba SO_4$) are less soluble in H_2O than uni – univalent electrolytes like KCl, similarly higher the distance between the ions, lesser is the force of attraction and hence more is the solubility. In the case of Li Cl or CsCl, the distance between Li^+ or Cl^- is smaller than Cs^+ or Cl^-. Hence the solubility of Li Cl is less than the solubility of CsCl in H_2O.

Non-Ionic Solids

Non – Ionic Solids are either polar or non polar. If the solute and the solvent have similar characters chemically, (i.e., both polar or non polar), the solubility is found to be low.

c) Nature of Solvents

Nature of solvent has a greater influence on the solubility of solutes. That is why, substances do have different solubility in different solvents. The force of attraction having a greater influence in solubility, is inversely proportional to the dielectric constant. It may be noted that higher the dielectric constant lower the force of attraction and hence higher is the solubility. If the dielectric constant is high, the substance is more soluble than in the solvent of low dielectric constant. For instance, NaCl dissolves more in water (D = 80) than in benzene (2.3). Thus, ionic solids dissolve to a larger extent in a solvent with high dielectric constant than in a solvent having a low dielectric constant.

Non-Ionic Solids

If the solvent is polar, it will dissolve polar solute and if it is non polar solvent it dissolves non polar solute. The underlying principle behind this process is 'like – dissolves – like'.

Henry's Law

In 1803, William Henry proposed a law which is now famously known as Henry's law. This law relates the solubility of a gas into a liquid to its pressure at a given temperature. Henry's law states that "the amount of a gas that dissolves in a given amount of a liquid at a constant temperature is proportional to the partial pressure of the gases present in the liquid provided, the gas does not undergo any chemical change during the formation of solutions".

In can be expressed mathematically as

$$M \propto P$$

$$\text{Or } M = K_H P$$

Where 'M' is the mass of the gas dissolved in a unit volume of the solvent. It is the pressure of the gas in equilibrium with the solvent and K_H is the Henry's constant which depends on the nature of the gas and solvent at given temperature.

Henery's law may also be noted as follows. At constant temperature, the solubility of a gas in a given liquid is directly proportional to the pressure of the gas in equilibrium with the liquid at that temperature.

It may be noted, that at a given temperature, if the mixture of gases are in simultaneous equilibrium with the liquid then, the solubility of any gas in the mixture is directly proportional to the partial pressure of that gas in the mixture. This is in accordance with the Dalton's law of partial pressures of a mixture of gases.

The mole fraction of the gas in the solution, and Henry's law can be related in mathematical form as -

$$K_A \propto P_A$$

$$\text{Or, } K_A = K_H P_A$$

Where P_A is the partial pressure of the gas being dissolved in the solution and K_H is the Henry's constant as stated above. The value of K_H, as mentioned above depends on the nature of the gas, solute and temperature. The units of K_H is atm^{-1} or bar^{-1}.

Sometimes the above equation is written as –

$$P_A \propto K_A$$

Or, $P_A = K_H x_A$

Here the unit will be atm or bar.

Graphical Representation of Henry's Law

It becomes convenient to draw a curve by plotting (using the above equation.) Mole fraction x_A verses corresponding equilibrium pressure (P_A). The plot is a straight line passing through the origin of the graph. The slope of the line gives the K_H (in atm. Or bar).

The plot shown in Fig. 6.2 is for solubility of HCl gas in Cyclohexane at 293 K.

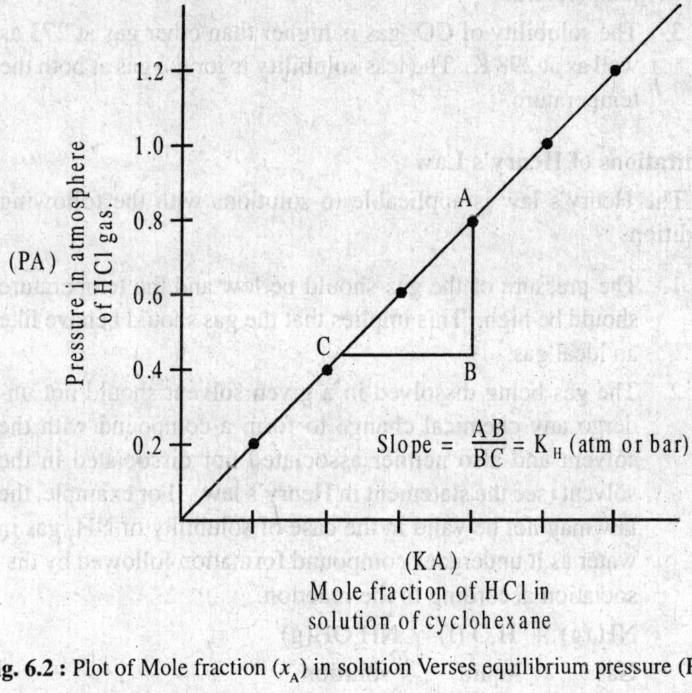

Fig. 6.2 : Plot of Mole fraction (x_A) in solution Verses equilibrium pressure (P_A)

Gas	*Temp (K)*	K_H *(atm-1)*	K_A *(bar-1)*
CO_2	273	1.4 x 10-3	1.38 x 10-3
	298	6.1 x 10-5	6.02 x 10-4
N_2	273	1.9 x 10-5	1.87 x 10-5
	298	1.2 x 10-5	1.18 x 10-5
O_2	273	4.5 x 10-5	4.44 x 10-5
	298	2.3×10^{-5}	2.27×10^{-5}
He	273	6.8×10^{-6}	6.7×10^{-6}

From those values the following points may be known –

1. Greater the value of K_H higher is the solubility of the gas.
2. The value of K_H decreases with increase in temperature and hence the solubility of the gas decreases with increase in temperature.
3. The solubility of CO_2 gas is higher than other gas at 273 as well as at 298 K. The less solubility is for the gas at both the temperature.

Limitations of Henry's Law

The Henry's law is applicable to solutions with the following conditions :-

1. The pressure of the gas should be low and the temperature should be high. This implies that the gas should behave like an ideal gas.
2. The gas being dissolved in a given solvent should not undergo any chemical change to form a compound with the solvent and also neither associated nor dissociated in the solvent (see the statement in Henry's law). For example, the law may not be valid in the case of solubility of NH_3 gas in water as it undergoes compound formation followed by dissociation according to the reaction.

   $NH_3(g) + H_2O(l)\ \gamma\ NH_4OH(g)$

   Gas liquid solution

   $NH_4OH + (aq) \rightleftharpoons NH_4^+(aq) + OH^-(aq)$

Similarly, the law may not be applicable for the dissolution of HCl gas in H_2O as solvent because it undergoes dissociation according to the reaction given by

$$\text{HCl (g)} + \text{H2O (aq)}\ \gamma\ \text{HCl aq}$$

$$\text{HCl (aq)}\ \gamma\ \text{H}^+\ \text{(aq)} + \text{Cl}^-\ \text{(aq)}$$

Application of Henry's Law

The Henry law of gas solubility play a greater role and premier role in soft drink industries.

1. In the Production of Carbonated Beverages

Carbon dioxide (CO_2) is dissolved under higher pressure in soft drinks, soda water as well as in bear, vodka etc. The containers of these drinks are well sealed and when the bottle or the container is opened (to air), the partial pressure of CO_2 above the solution decreases and blashes of CO_2 released.

2. Deep-Sea Divers

Deep sea divers relay completely upon the compressed air for their oxygen demand. According to Henry's law, solubilities of gases increase with pressure. Thus, both N_2 and O_2 dissolve considerably in the blood and other body fluids. The dissolved oxygen is used for the metabolism. But N_2, because of higher solubilities remain as dissolved in the blood as well as other body fluids. It forms bubbles when the sea – divers come to the atmospheric pressure. As the sea divers comes out, they suffer from a terrific unbearable pain from a disease called bends or decompression sickness due to the formation of bubbles of N_2 in the blood as well as other body fluids. Therefore, N_2 is replaced by 'He', an inert gas which is more soluble than N_2 in the biological fluids. Due to this reason a mixture of 2% O_2 and 90%He is used to avoid bend which is greater after in impresses of the nerves.

3. In the Function of Lungs

When air is breathed in, oxygen content is high in the lungs. The oxygen combines with haemoglobin to form oxyhaemoglobin. In the tissues the partial pressures of O_2 is low and hence O_2 is released from hemoglobin which is utilized for the biochemical function of the cells.

Colligative Properties

Certain definite properties of ideal solution depend only on the number of the particles of the solute (in the form of molecules or ions) in a given amount of the solvent and do not depend on the nature of the solute. Such properties are usually known as colligative properties.

The colligative properties of some system are –

1. Relative lowering of vapour pressure
2. Osmotic pressure
3. Elevation in boiling point
4. Depression in freezing point.

1. Relative Lowering Of Vapour Pressure (Rlv)

According to Raoult's law, the relative vapour pressure of a solution is lower than that of the vapour pressure of pure solvent.

The relative lowering of vapour pressure is given by an expression.

$$\frac{P^o - P_s}{P^o} = \frac{n_1}{n_1 + n_2}$$

Where P^0 is the vapour pressure of pure solvent.
P_s is the vapour pressure of the solution.
n_1 is the number of moles of solvent
n_2 is the number of moles of solute.

The conclusion is that the relative lowering of vapour pressure is a colligative properties which depends only on the number of solute particles and not on the nature of the solute.

Molecular masses of the solutes can be determined (experimentally) from their lowering of vapour pressure values, using Raoult's law.

$$\frac{P^0 - P_s}{P^0} = \frac{n_2}{n_1 + n_2} \text{ ---- (1)}$$

Let us consider that W_1 gram solvent contains W_2 gram of the solute. Let M_1 and M_2 be the molar masses of the solvent and solute respectively, then we have

$$n_2 = \frac{W_2}{M_2} \text{ and } n_1 = \frac{W_1}{M_1}$$

On substituting these values in the above expression (equation - 1), we get

$$\frac{P^0 - P_s}{P^0} = \frac{n_2}{n_1 + n_2} = \left(\frac{W_2 / M_2}{W_1 / M_1 + \frac{W_2}{M_2}} = \frac{W_2 \ M_1}{M_2 \ W_1} \right)$$

$$\therefore \left(\frac{P^0 - P_s}{P^0} \right) = \left(\frac{W_2 \ M_1}{W_1 \ M_2} \right)$$

For a dilute solution, n_2 can be neglected in comparison with n_1 (large quantity of water) so that Raoult's law can be written as

$$\frac{P^0 - P_s}{P^0} = \frac{n_2}{n_1} \text{ ----- (2)}$$

$$\text{then, } \frac{P^0 - P_s}{P^0} = \frac{W_2 / M_2}{W_1 / M_1} = \frac{W_2 \ M_1}{W_1 M_2}$$

$$\therefore \frac{P^0 - P_s}{P^0} = \frac{W_2 \ M_1}{W_1 M_2} \text{ ----- (3)}$$

This equation is used to calculate the molecular mass of the non-volatile solute provided the values of vapour pressure of the pure solute (P_0) and that of the solution P_s are known. Experimentally knowing the exact amount of the solute diffuse in a definite amount of the solute, then the molecular mass M_2 of the solute can be calculated provided M_1 is known. This method, have been preferred over other methods because elevation in boiling point or depression in freezing point, give result accurately and easily.

2. Osmotic Pressure

There is a natural tendency that the solutes in a solution to diffuse from higher concentration to a lower concentration to attain uniform distribution throughout the solution.

Therefore, this phenomenon of the movement of particles of a solute from a more concentrated to less concentrated solution so as to attain uniform concentration throughout the solution is known as diffusion.

The above phonomenon can be explained by a simple experiment. Suppose a concentrated solution of $CuSO_4$ (deep blue colour) is placed in a beaker and water (or dilute $CuSO_4$ solun) is added very slowly to the beaker without disturbing it. Well defined two layers are formed. If the beaker is allowed for a longer time the solution in the beaker becomes uniformly blue in colour throughout. Thus, obviously, due to the presence of Cu^{2+} and SO_4^{2-} which move slowly into the solvent and the solvent molecules moves into the copper sulphate solution. In other words, the solute and solvent particles mix spontaneously with each other and thereby creating an uniformity in the concentration of the solution.

The same experiment can be performed in a slighter diffuse manner by using a semi permeable membrane. Suppose beaker is divided into two components by semi permeable membrane which allows the solvent particles to pass through in and out of the membrane. Suppose if we take $CuSO_4$ solution in one component and water into the other, it is observed that the level of the solution begins to rise. This is due to the fact that greater number of solute molecules puts into the solution side through semi permeable membrane. Similarly flow of solvent molecules from the dilute solution to the concentrated solution through semi permeable membrane till the concentration on both sides of semi permeable membrane becomes equal.

$CuSO_4$ solution.

a)

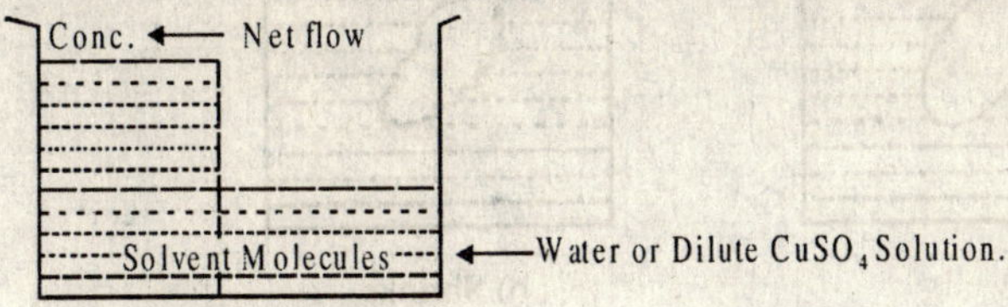

b)

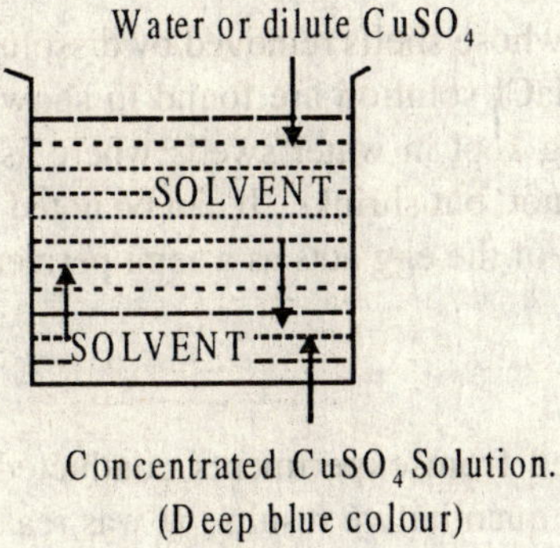

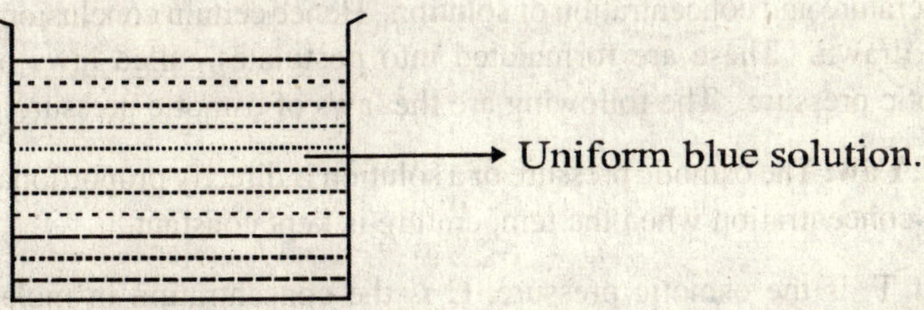

OSMOSIS

The non spontaneous flow of the solvent molecules from the solvent to the solution or from a less concentrated solution to a more concentrated solution through a semi permeable membrane is called Osmosis'. (Greek: Push)

The Phenomenon of osmosis can be demonstrated as shown in the following diagram

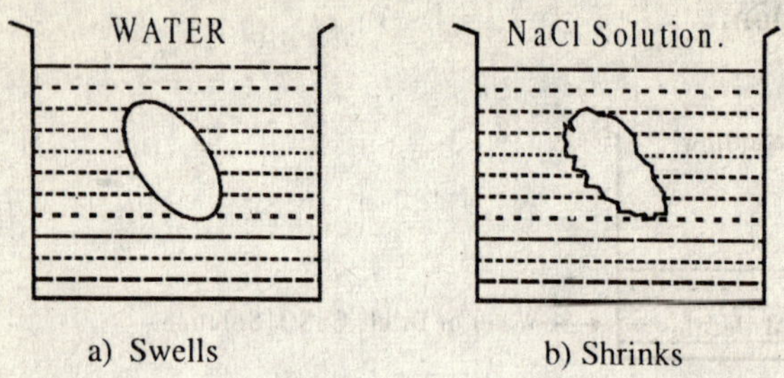

Fig. 6.3: Egg without outer shell.

When two eggs seperately, whose shells removed by dissolution in HCl, kept in water and in NaCl solution are found to show the following observation. The egg kept in water swells where as the one kept in NaCl solution does not, but shrinks. It can be noted that the membrane beneath the shell of the egg acts as a semi permeable membrane.

Laws of Osmotic Presssure

Experimental data collected from experiments conducted by Pfettor and others led to certain quantitative results. It was realised that the osmotic pressure of solutions depends upon both the temperature and concentration of solution. Hence certain conclusions were drawn. These are formulated into postulates called laws of osmotic pressure. The following are the laws of osmotic pressure.

I. Law: The osmotic pressure of a solution is directly proportional to its concentration when the temperature is kept constant.

If T_1 is the osmotic pressure, C is the concentration in molar units, then at constant temperature

$$T_1 \propto c;$$

$$T_1 = \text{Constant} \times C$$

$$\text{or } \frac{T_1}{C} = \text{Constant} \text{ ———— (1)}$$

If 1 g mole of the solute is present in V litres of the solution then

$$C = \frac{1}{V}$$

$\therefore$ T_1 V = constant at constant temperature. Vant Hoff pointed out that Osmotic pressure obeys Boyle's law

II. Law: The osmotic pressure (T_1) of a solution varies directly to the absolute temperature T. If the concentration of the solution remaining constant.

$$T_1 \propto c;\ T_1 = \text{Constant} \times C$$

$$\text{or } \frac{T_1}{C} = \text{Constant} \text{———————} (2)$$

This is analogous to Charle's law of gases. The results of these two laws can be combined to get the result.

$T_1V = RT$ ——————(3) for dilute dilutions. Where R is constant. This is analogous to ideal gas equation. PV=RT for 1 mole of the gas.

It is interesting that by inserting actual values of Osmotic pressure for a definite dilute solution of known concentration at a given temperature, the value of R in the above equation (3) is found to be 2.147 which is almost identical with the gas constant R.

The equation (3) is referred to as the Van't Hoff's equation for dilute solutions, which is exactly analogous to the ideal gas equation. Thus, it is clear, that Van't Hoff(1885) put forth his theroy of solutions in the following words:

"A substance in the solution behaves exactly like a gas and the osmotic pressure of a dilute solution is equal to the pressure, which the solute would exert if it were a gas at the same temperature and occupying the same volume as the solution.

Measurement of Osmotic Pressure

Measurement of osmotic pressure of a dilute solution is difficult task since it requires a satisfactory semi permeable membrane. However, in semi permeable membrane is prepared, osmotic pressure of a dilute solution can be determined / measured experimentally. Many methods have been designed for measuring osmotic pressure. However, one such method is explained below.

Berkeley and Hartley's method.

The basic principle involved in this method is that the solvent is not allowed to enter the solution by applying an external pressure just enough to prevent the flow of the solvent through the membrane. This applied pressure is the osmotic pressure of the solution. This method is superior to P fetter's method.

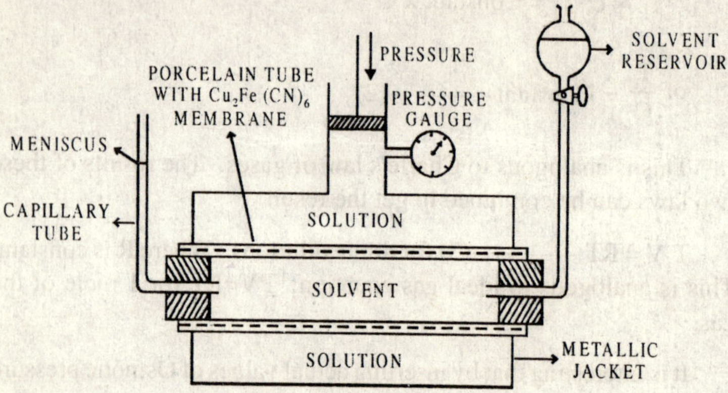

Berkeley and Hartley's apparatus

It consists of porous tube which opens at its either end. The tube contains semi -permeable membrane made of copper ferrocynide. The tube is connected to reservoir on one side and on opposite of this reservior another tube T is filled to the porous tube.

The porous tube filled with a pure solvent (H_2O) so that, the level in the tube T stands at the mark M as shown in fig 6.3. The porous tube is fitted in to an outer vessel made of gun metal. At the top of the vessel a wide tube is fitted and at the top of which the frictionless piston and pressure gauge is fitted as shown in fig. The solution whose osmotic pressure has to be determined is taken in the outer vessel. Due to osmosis, the level q, of solvent in the tube T falls down Now, the pressure is applied on the solution side by means of the piston to keep the level in the tube T at M as before. The pressure applied to just maintain this condition is a measure of osmotic pressure which can be read directly on the pressure gauge.

This Berkeley's and Hartley's method is superior to other methods for the following reasons.

1) During the measurement of osmotic pressure of a given solution, the semi permeable membrane is not strained since the osmotic pressure is carefully balanced by the external pressure.
2) The concentration of the solution does not alter because the entry of the solvent molecules into the solution is prevented by the external pressure.
3) This method is quick and easy since the time taken for the measurement of osmotic pressure is less compared to other methods.

Isotonic Solution:

Those solutions having the same osmotic pressure at the same temperature are called isotonic solutions.

This is evident from the equation $PV = cRT$ if two solutions have the same values of c and T, then they must have the some value of P. In other words, solutions of equimolar concentration at the same temperature must have the same osmotic pressure. It is also evident from the above equation that, the osmotic pressure depends on the number of moles of the solute per liter of the solution and not on the nature of the solute. Hence, osmotic pressure is a colligative property.

Biological importance of osmosis

The phenomenon of osmosis plays a vital role in plant and biological systems. The growth of plant and animals is largely depend on the osmosis. It is known that, the plants and animal systems (bodies) are composed of cells. The cells contain fluid called cell sap and its wall is composed of cytoplasmic membrane which is semipermeability in nature. This semipermeable is responsible for the phenomenon of osmosis in living organisms and plants as well.

Hemolysis: If a living cell comes in contact with water (or dilute solutions), there is a tendency for water molecules to enter the cell through the cell wall. The cell swells up and cell raptures. The process of rupturing of the cell is known as Hemolysis. This is mainly due to the osmotic pressure difference. Since the osmotic pressure of water (or dilute solution) is less than that of the cell sap and hence there is a greater tendency for water molecules to enter into the cell (Fig – 2)

Plasmolysis : If the cell comes in contact with a solution having osmotic pressure, then the cell starts shrinking due to outgoing of water molecules from the cell through the cell wal! with the solution side. The phenomemen of shrinking of the cell is called plasmolysis.

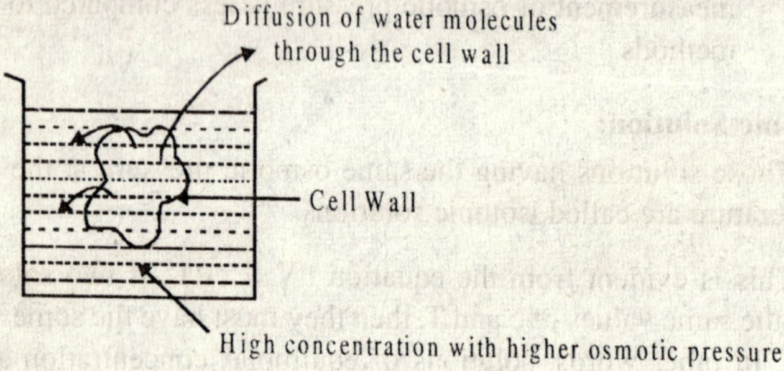

Fig.6.4: Schematic representation of the phenomenon of plasmolysis.

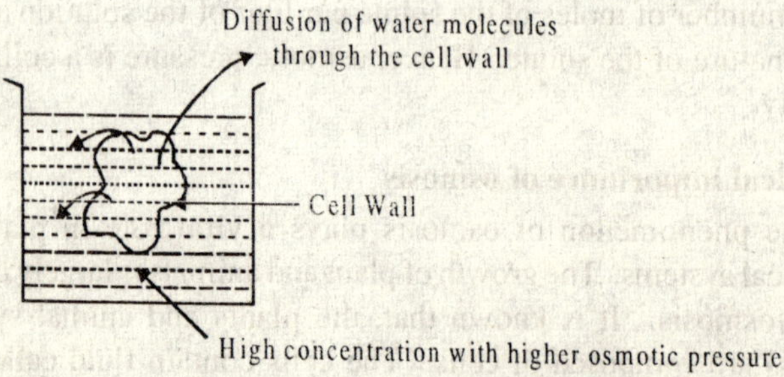

Fig-6.5 : Swelling / Repturing of the cell leading to hemolysis.

Some other important process regulated by osmosis.

1) The osmosis regulates certain process in plants and animals are given below:

a) Water absorbed by the plants from the soil. Water from the soil raises up the through roots of plants due to osmosis. This is due to the fact that, the concentration of the cell sap inside the root hair cell's is higher than that of the water in the soil. The osmotic pressure is also different which causes upward movement of water molecules even to the top of a tall tree. It may be noted that hair root cells acts as capillaries through which water moves up.

b) In animals, the movement of water molecules to the different parts of the body is due to osmosis.

c) Stretching of leaves or spring up of bud flower is due to osmosis which is the phenomenon.

d) Osmosis also plays an important role in growth of the plants and germination of seeds.

e) Bursting of red blood cells when placed in water or dilute solution is caused by osmosis. The some phenomenon of osmosis causes certain actions like opening and closing of flowers of certain plants.

f) It is interesting to note that the concentration of different species present in the blood plasma is approximately equal to 0.91% aqueous solution of NaCl by mass. Hence, 0.91% solution of NaCl is isotonic with human red blood cells (RBC) so that RBC undergo swelling not plasmolysis. A solution of more than 0.91% NaCl is called hypertonic solution. In this solution RBC (placed) undergo plasmolysis. Similar solution of NaCl less than 0.91% is called hypotonic solution. In this solution the RBC swells up and even they surge up. Thus it becomes essential that the solution flow into the blood stream should have the some osmotic pressure equal on outside or inside of the cell is created by transport of Na^+ and Cl^- ions across the cell membrane. It may be noted that the biological cells allow the passage of not only water molecules but also to other selected matter called metabolites across their membranes. This condition helps the entry of the nutrients and disposal of the waste materials. This active transport takes place across the membrane from a region of higher concentration to a region of lower concentration.

Depression in freezing point:

General discussion: Freezing point of a substance is the temperature at which the soild and the liquid forms of the substance are in equilibrium, i.e., the solid and the liquid forms of substance have the same vapour pressure. It is observed that the freezing point of the solution is always lower than that of the pure solvent. The decrease is called the depression in freezing point. The reason for the depression can be explained as follows.

It was found that the vapour pressure of the solution is less than that of the pure solvent. As freezing point is the temperature at which the vapour pressure of the liquid and the solid phase are equal, therefore, for the solution, this will occur at a lower temperature (because lower the temperature, lower is the vapour pressure).

Alternatively, the depression in freezing point may be explained on the basis of plots of vapour pressure verses temperature.

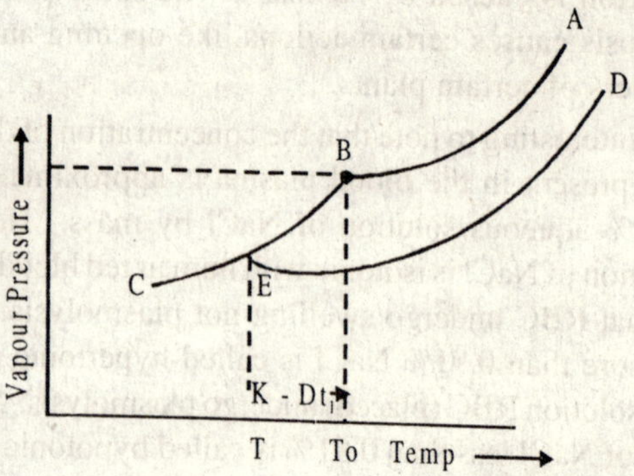

On observing the curves, the vapour pressure of the liquid solvent decreases along the curve AB, at 'B', the solid starts appearing and the vapour pressure decreases steeply along the path BC (because solids have lower vapour pressure). At B, the liquid and the solid solvent are in equilibrium and have the same vapour pressure.

Thus B represents the freezing point of the pure solvent. As at any temperature, the vapour pressure of solution is less than that of the solvent, the curve for the solution lies below that of the solvent. Therefore it will follow the path DE. At E the solid appears. Hence, E represents the freezing point T of the solution. Obviously, T is less than T_0. The difference, is called the depression in freezing point, DT_f which is given by $DT_f = T_0 - T$.

The depression in freezing point depends on the concentration of the solute in a solution and has been found to be related to molality shown as below.

$$DT_f = K_f \text{ x } m.$$

Where Kf : is a constant known as molar depression constant or cryoscopic constant of the solvent and m is the molality of the solution n is the number of moles of the solute dissolved in 1000 gram of the solvent.

If m = 1, $DT_f = K_f$ hence, the expression, $DT_f = K_f$ x m can be derived in a simple manner exactly in the same way as in case of elevation in boiling point.

$$DT_f = K_f m$$

Where DT_f = depression in freezing point = $T_0 - T_s$.

M = Molality of the solution

K_f = molal depression constant or cryoscopic constant in K/m or k kg / mole.

$$DT_f = \frac{1000 \, k_f \, w_2}{w_1 \, M_2} \text{ or } M_2 = \frac{1000 \, k_f \, w_2}{w_1 \, DT_f}$$

Where M_2 = molecular mass of the solute.

W_2 = weight of the solute in grams.

W_1 = weight of the solvent in grams.

Elevation in boiling point:

General discussion: The boiling point of a liquid is the temperature at which the vapour pressure of the liquid becomes equal to the atmospheric pressure.

For example, vapour pressure of water at 100^0 C (373k) is 1.013 bar (= 1atm). This is the reason for the water to boils at 373K because this temperature its vapour pressure becomes equal to one atmospheric pressure.

It is found that, the boiling point of the solution is always higher than that of the pure solvent and vapoure pressure increases with increases in temperature. Hence the solution has to be heated more to make the vapour pressure equal to the atmospheric pressure.

Alternatively, the elevation in boiling point may be explained on the basis of the plots of vapour pressure versus temperature as follows.

Vapour pressure of the solvent increases with increases in temperature as shown by the curve AB. At any temperature, vapour pressure of the solution is less than that of the solvent. The curve for the solution lies below than that of the solvent as shown by the curve CD. The temperatures at which the vapour pressure of the solvent and the solution becomes equal to the atmospheric pressure are T_0 and T_1 respectively. obviously $T_1 > T_0$. The difference is called the elevation in boiling point DT_b which is given by $DT_b = T_1 - T_0$.

The elevation in boiling point depends upon concentration of the solute in a solution and has found to be related with molality 'm' as below.

$$DT_b = K_b \times m.$$

Where K_b is called the molal elevation constant or ebullioscopic constant and 'm' is the molality of the solution.

If $m = 1$ and $DT_b = K_b$ Hence,

Molal elevation constant may be defined as the elevation in boiling point when the molality of the solution is unity, i.e., one mole of the solute is dissolved in 1 kg (1000g) of the solvent. The units of K_b are, therefore, degree/molality is k/m or c/m or K kg/mol.

The molal elevation constants (ebullioscopic constants) of a few solvents are given below.

Solvent	K_b *(KKg mol^{-1})*
Water	0.52
Benzene	2.53
[illegible]	1.20

Carbon disulphide	2.34
Chloroform	3.63
Cyclohexane	2.79
Carbon tetrachloride	5.03
Ether	2.02
Acetone	1.72
Acetic acid	2.93

Derivation of the expression DTb = Kb x m.

The above relationship may be derived in a simple manner as follows.

It is evident that greater the lowering of vapour pressure (D_p), higher is the elevation in boiling point (DT_b)

$DT_b \propto D_p$.

But according to the Raoult's law, $DT \propto X_2$, the mole fraction of the solute in the solution.

Hence $DT_b \propto X_2$, or $DT_b = kX_2$.

Where k is a constant of proportionality.

But $X_2 = \frac{n_2}{n_1 + n_2} = \frac{n_2}{n_1}$ if the solution is dilute.

is $X_2 = \frac{n_2}{w_1 / m_1}$ $\therefore DT_b = XKM_1 \frac{n_2}{w_1}$

If the weight of solvent W_1 = 1kg, then evidently, molality of thc solution. Also for a given solvent, its molar mass m is constant so that $km_1 = k_b$, another constant. Hence the above result reduces to $DT_b = K_b m$.

Calculation of molecular mass of the solute:

As molality is the number of moles of the solute dissolved per 1000 kg of the solvent. If W_2 grams of the solute of molecular mass m_2 are dissolved in W1 grams of the solvent.

$$m = \frac{W_2}{W_2} \times \frac{1000}{W_1}$$

Hence, the above formula becomes

$$DT_b = K_b \text{ x } \frac{W_2}{M_2} \text{ x } \frac{1000}{W_1}$$

$$M_2 = \frac{1000 \text{ x } K_b \text{ x } W_2}{DT_b \text{ x } W_1}$$

OR

This formula is often used for the calculation molecular masses of non ionic solutes.

QUESTIONS CARRYING ONE MARKS

1) Define Molarity and Molality.
2) Define normality and mole fraction.
3) Give an example of a solution containing a liquid solute in a solid solution.
4) When and why molality preferred over molarity.
5) What is the effect of temperature on molarity of a solution.
6) Why the vapour pressure of a liquid is constant at constant temperature?
7) Two liquids A and B are mixed the solution is found to be cooler. What do you conclude?
8) Define the term freezing point.
9) Write Vant – hoff's equation.
10) Calculate osmotic pressure of 5% solution of glucose.
11) Explain the term "Osmotic Pressure".
12) What is meant by isotonic solution.
13) State Roult's law.
14) What is meant by Osmosis.

QUESTIONS CARRYING 3 OR 4 MARKS

1) Calculate the mole fraction of ethanol and water in a sample of rectified spirit which contains 95% ethanol by weight. (0.88, 0.12)
2) Find the molarity and molality of a 15% solution of $H_2 SO_4$ (density of $H_2 SO_4$ = 1.020 gcm^{-3}. (1.56M)
3) A sugar syrap of weight 214.2 g contains 34.2 g of sugar ($C_6 H_{12} O_6$). Calculate (i) Molal concentration (ii) Mole fraction of sugar.
4) The boiling point of water becomes 100.052°C. If 1.5g of a non – volatile solute is dissolved in 100ml of it calculate the molecular weight of the solute.
(Kb for H_2O = 0.52 kg mol^{-1})
5) Explain why normality and molarity change with temperature but molality and mole fraction does not.
6) What are azeotrophic mixtures? Give examples.
7) What do you understand by the term solid solution?
8) State and explain Henry's law. Show that volume of a gas dissolved in a given volume of a solvent is independent of pressure of the gas.
9) Give reason for the following.
   a. On opening a carbonated cold drink bottle, bubbles of gas comes out.
   b. On adding common salt to a bottle of carbonated cold drink froth comes out.
10) What do you understand by the term "solid solution"? Give example? Give an example.
11) Explain the following.
    a. Solubility of a gas in a given liquid decreases with raise in temperature.
    b. Mixture of alcohol and water cannot be separated by fractional distillation.
12) What is the Molarity and Molality of a 13% solution of H_2SO_4? It's density is 1.09 g/ml.

13) Explain the term lowering of vapour pressure and relative lowering of vapour pressure what are their units?

14) Discuss vanti-Hoffi theory of dilute solutions. What is Vant Hoff's factor?

15) What certain amount of solute is added to 100g of water at 25°C, the vapour pressure reduces to one half of that for pure water. The vapour pressure of water is 23.76mm Hg. Find the amount of salt added.

16) a. State Roult's law.

    b. A solution of 5.85g of Sodium nitrate in 100g of water freezes at - 3.04°C. Calculate the molecular mass of sodium nitrate and account for the abnormal value (Kf for water = 1.86 K mol^{-1})

17) What are colligative properties and give examples of each property.

18) State Roult's law of elevation of boiling point of solvent.

19) State and explain Raoults law. Show graphically the variation of total vapour pressure over mixture of two volatile liquids with the composition of mixture.

20) State and explain Henry's law. What are its limitations?

21) What are ideal and nonideal solutions? What type of non idealities are exhibited by cyclohexane, ethanol and cydohexane chloroform mixtures. Give reasons for your answer.

22) Define the solubility of a solid in a liquid. Briefly describe the various factors on that the solubility of a solid in a liquid depends.

7

ELECROLYTIC DISSOCIATION AND MASS LAW

Electrochemistry is a branch of chemistry which deals with the study of two types of chemical reactions.

a) Which occurs themselves with the release of electrical energy

b) The reaction which takes place only by the supply of electrical energy.

Based on the electrical conducting capacity of the substance they are divided into two types

1) Insulators, which does not conduct electricity. Example: plastic glass, diamond etc.
2) Conductors, which allows electric current to pass through them. Example: Graphite, all metals etc.

The conductors which allows electric current to pass through them is again classified into two types based on the nature of the charge carriers. They are.

(a) Electric conductors:	**(b) Electrolytic conductors:**
1. The charge carriers are electrons.	1. The charge carriers are ions.
2. There is no transfer of matter.	2. There is a transfer of matter
3. There is no charge in the chemical property of a conductor.	3. There is a change in the chemical property of a conductor and the chemical changes occur on the surface of the electrode.
4. With the increase in temperature, the conductivity decreases due to increase in resistance.	4. With increase in temperature, conductivity increases due to decrease in resistance.

Electrolyte	Non – electrolyte
An electrolyte is a substance which is capable of producing ions when dissolved in water or in molten state.Example: All acids, all bases and almost all salts.	A non – electrolyte is a substance which cannot produce ions in aqueous solution or in molten state.Example: sugar, urea, Benzene, CCl4 etc.
Strong electrolyte	**Weak electrolyte**
An electrolyte which undergoes complete dissociation is called strong electrolyte.Example: All strong acids, strong bases and almost all salts.	An electrolyte which undergoes partial dissociation is called a weak electrolyteExample: All weak acids and weak bases.

Conductance: When a cell dipped in an electrolyte solution, the resistance of the solution is directly proportional to the distance between the electrodes and inversely proportional to area of cross-section of the electrode.

$$\text{ie, } R \propto \frac{\ell}{a}$$

The reciprocal of resistance is called conductance.

$$\therefore \quad \frac{1}{R} \propto \frac{a}{l}$$

i.e., conductance of an electrolyte solution is directly proportional to area of cross -section of electrode and inversely proportional to distance between the electrodes.

$\frac{1}{R} = K.\frac{\ell}{a}$ Where K is proportionality constant which is also called specific conductance

$$K = \frac{\ell}{Ra}$$

$$K = \frac{1}{R}.\frac{\ell}{a}$$

For a given cell, the distance between the electrodes (ℓ) is constant and area is also constant. Therefore, $\frac{\ell}{\alpha}$ is again a constant known as cell constant.

$$\therefore \; C_P = \frac{\ell}{a}$$

$$\therefore \; K = C \times \frac{1}{R}$$

$$\therefore \text{ Specific conductance} = \frac{\text{cell constant}}{\text{Resistance}}$$

Specific conductance [K Ω^{-1} m^{-1}] : specific conductance is defined as "the conductance of 1 m^3 solution of an electrolyte increased by using electrodes of each area 1 m^2 which are separated by a distance of one m. It is denoted by a symbol 'K'. The unit is given by Sm^{-1} [siemens per meter].

In conductrometric titrations, the end point is defined by measuring the variations of conductance, without using an indicator. The conductivity of all the electrolytic solution depends on the following two factors.

1) Number of ions present in the solution
2) Mobility of the ions.

Conductometric titration of a strong acid versus a strong base:

The conductivity cell is washed with distilled H_2O. A volume of standard HCl is taken in the conductivity cell. If the electrodes are not completely dipped, add distilled H_2O. The cell is connected to one arm of the meter bridge and a suitable resistance is applied in the resistance box.

According to the principle of Meter Bridge

$$\frac{R_{Hcl}}{R_{base}} = \frac{\ell}{100-\ell}$$

Hence, the resistance of HCl solution can be calculated. The reciprocal of resistance gives conductance.

ie, $R = \frac{1}{C}$ (or) $\frac{1}{R} = C$

The conducto metric titration is carried out as follows. The NaOH solution which is under examination is added from burette in small proportion to HCl Solution. After each addition, the solution is stirred well and the resistance is determined.

In the beginning conductivity of the solution goes on increasing, but on adding NaOH, a fast moving H^+ ion is replaced by a slow moving Na^+ ion. After equivalent point, further addition of NaOH contributes number of Na^+ and OH^- ions. As a result, the conductivity of the solution increases. Plot a graph of conductivity verses volume of NaOH. It can be calculated from the graph, that the volume of NaOH required to react with known volume of standard HCl. By knowing normality, volume of HCl and volume of NaOH,. Normality of NaOH can be calculated

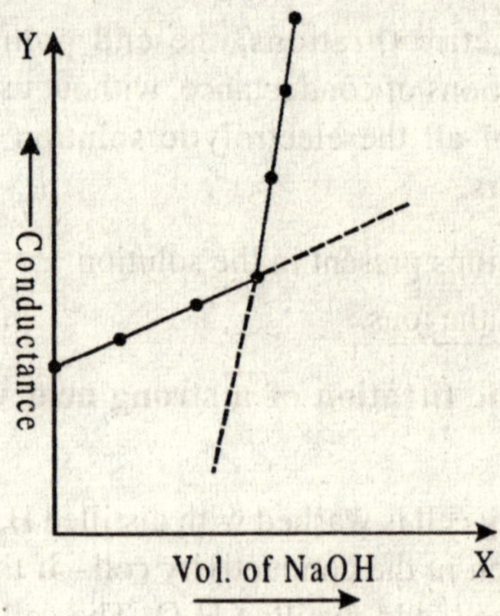

weak acid and strong base

$$N_{NaOH} = \frac{\text{Normality} \times \text{Volume of HCl}}{\text{Volume of NaOH.}}$$

In conductometric titration of weak acid versus a strong base the conductance of the solution is small in the beginning. Because the acid is weak which dissociates partially and contributed a few ions. Addition of NaOH results in the formation of a salt which is strong electrolyte, which produces more number of ions. As a result, the conductance goes on increasing gradually. After equivalence point, further addition of NaOH increases the conductance rapidly by contributing more number of ions.

A graph of conductance verses volume of NaOH solution is given by.

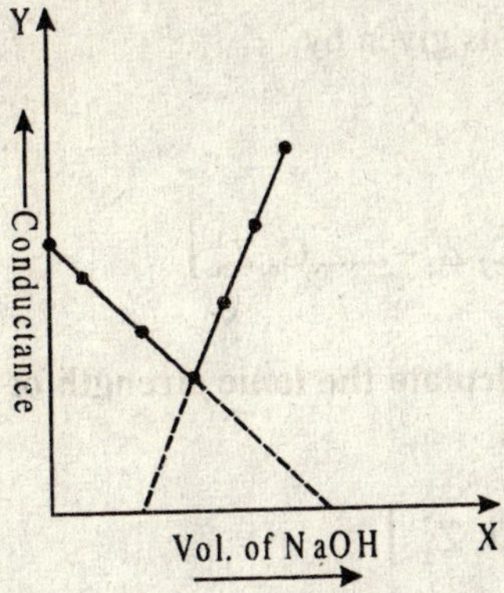

strong acid and strong base

Activity and activity co-efficient

A solution of strong electrolyte will have two different concentration due to the fact that, an ion in a solution is not exact in its full individual effect towards the conductance, the observed activity is called effective concentration which is different from actual concentration. The effective concentration is also called, activity represented by a symbol 'á'. The actual concentration is presented as 'c'. The ratio of the effective concentration to the actual concentration is called activity coefficient and is denoted as 'υ'

$$\gamma = \frac{\alpha}{c}$$

Debye and Huckel derived an equation ie,

$\log \gamma = -A Z^{+} Z^{-} \sqrt{I}$ where I is its lonic strength

Where A is Constant which depends on the nature of solvent and temperature.

A Z^+ Z^- represent – The charges of cation and anions respectively neglecting the sign.

I – Ionic strength.

Ionic strength is given by

$$I = \frac{1}{2} \Sigma C_i Z_i^2$$

$$I = \frac{1}{2}\left[C_1 Z_1^2 + C_2 Z_2^2 + \ldots\ldots\ldots C_n Z_n^2\right]$$

Example – 1: Calculate the ionic strength of 0.04 molar copper sulphate solution.

$$I = \frac{1}{2}\left[C_1 Z_1^2 + C_2 Z_2^2\right]$$

$$= \frac{1}{2}\left[0.04 \times 4 + 0.04 \times 4\right]$$

$$= 0.16$$

Example – 2: Calculate the activity co-efficient of 0.01 M HCl solution given A = 0.509

$HCl \rightarrow H^+ + Cl^-$

0.01 0.01 0.01

$$= \frac{1}{2}\left[0.01 \times 1 + 0.01 \times 1\right]$$

$$I = 0.01$$

$$\log \gamma = -A\, z^+ z^- \sqrt{I}$$

$$\log \gamma = -(0.509)\,(1)(1)\,\sqrt{0.01}$$

$$\log \gamma = -0.0509$$

$$\log \gamma = \text{antilong } -(0.509)$$

$$\log \gamma = 0.8894$$

Common ion effect:

Consider the dissociation of weak electrolyte such as acetic acid.

$$CH_3COOH \rightarrow H^+_{(aq)} + CH_3COO^-_{(aq)}$$

Applying law of mass action to the above equilibrium

$$K_a = \frac{[CH_3COO^-][H^+]}{[CH_3COOH]}$$

K_a is called the dissociation constant of the acid which is a constant. If a strong electrolyte like sodium acetate is added to acetic acid, there will be an increase in concentration of acetate ions by dissociating completely. Since Ka is constant, an increase in the concentration of acetate ions should accompany a corresponding decrease in the concentration of H^+ ions. The addition of acetate ions to acetic acid will shift the equilibrium to the left, thus, the addition of an electrolyte containing a common ion decreases the degree of dissociation of the weak acid.

Consider the dissociation of ammonium hydroxide,

$$NH_4OH_{(aq)} \rightarrow NH^+_{4\,(aq)} + OH^-_{aq}$$

Apply the law of mass action to the equilibrium

$$K_b = \frac{[NH_4^+][OH^-]}{[NH_4OH]}$$

K_b is known as dissociation constant of the base. If a strong electrolyte like ammonium chloride is added to ammonium hydroxide ions, there will be an increase in the concentration of ammonium ions by the complete dissociation of ammonium chloride. Since K_b is constant, an increase in concentration of ammonium ions should accompany a corresponding decrease in the concentration of OH^- ions. The addition of strong electrolyte containing a common ion to a solution of a weak base decreases the degree of dissociation of the base.

The effect of decrease in degree of dissociation of a weak electrolyte by the addition of another electrolyte containing a common ion is called common ion effect.

Solubility product:

When a solid solute dissolves in a liquid solvent there exists an equilibrium between the solution and the undissolved solid. Some of the solutes dissolve to a very large extent where as some others show only a slight solubility. The solute that dissolves very slightly is referred to as sparingly soluble salts.

Consider saturated solution of sparingly soluble salts like silver chloride, Bismuth sulphide etc.

$$AgCl \rightarrow Ag^{+}_{(aq)} + Cl^{-}_{(aq)}$$

Applying law of mass action.

$$K = \frac{[Ag^{+}][Cl^{-}]}{[AgCl]}$$

Concentration of AgCl (s) is constant at a given temperature.

$$K = \frac{[Ag^{+}][Cl^{-}]}{Constant}$$

$K_{sp} \rightarrow$ sp is alled the solubility product of the sparingly soluble salts.

In case of bismuth sulphide, the salt is dissociated to form Bi^{3+} and S^{2-} ions in the aqueous solution.

$$Bi_2S_3(s) \rightarrow 2Bi^{3+}_{(aq)} + 3S^{-2}_{(aq)}$$

$$K_{sp} \text{ of } Bi_2S_3 = [Bi^{3+}]^2 + [S^{-2}]^3$$

In general, Sparingly soluble salt A_xB_y, the equilibrium can be written as

$$A_x B_{y\,(s)} \rightarrow xA^{m+}_{(aq)} + y\,B^{n-}_{(aq)}$$

The solubility product principle may be stated as: "when a solution is saturated with a sparingly soluble ions raised to the appropriate powers is a constant value".

An important application of the solubility product principle is in the study of precipitation reaction. The condition of precipitation is

$$[A^{m+}]^x [B^{n+}]^y < K_{sp}$$

If the product of molar concentrations is less than K_{sp} , no precipitation occurs.

$$[A^{m+}]^x [B^{n-}]^y < K_{sp} \text{ No precipitate.}$$

Problems

1.Calculate the solubility of ZnS, if K_{sp} of the salt as 1.6×10^{-24} at 298K.

The dissociation of Zns in aqueous solution is given as,

$$Zns_{(s)} \rightarrow Zn^{2+}{}_{(aq)} + S^{-2}_{(aq)}$$

Let the solubility of ZnS be x.

Every soluble moleule of ZnS will give rise to one Zn^{2+} and one S^{2-}. Therefore,

Solubility of ZnS = $[Zn^{2+}] = [S^{2-}] = x$

$$[Zn^{2+}]\,[S^{2-}] = K_{sp} \text{ of Zns}$$

since $[Zn^{2+}] = [S^{2-}]$, the above equation may bc

written as

$$[Zn^{2+}] = [Zn^{2+}] = K_{sp} \text{ of ZnS}$$

$$x^2 = K_{sp}$$

$$\text{Solubility of ZnS} = x = \sqrt{K_{sp}}$$

$$=\sqrt{1.6\times10^{-24}\left(\text{mol dm}^{-3}\right)^2}$$

$$= 1.27 \times 10^{-12} \text{ mol dm}^{-3}$$

Electrochemical cells:

Electrochemical cell is a device which can convert chemical energy into electrical energy. It consists of two half cells referred to as single electrodes. The electrode potential arising at any half cell is expressed in terms of its reduction potential. When two half cells are connected, the one with the higher reduction potential acquires electron from the half cell having the lower reduction potential. The measured cell potential is the magnitude of the difference between the reduction potential of the half cells. The electrochemical cell can also be called as voltaic cell or galvanic cell. The reaction in which loss of electron takes place is referred as oxidation reaction and the reaction in which gain of electron takes place is referred as reduction reaction. The electrode at which oxidation takes place serves as a -ve electrode which supplies electrons, the electrode at which reduction takes place is a+ ve electrode which accepts electron The two half cells are connected externally through an ammeter. Then the electrolytes of a electrodes are connected through a salt bridge. The salt bridge is an inverted "U" – shaped tube containing KCl solution mixed with a solidifying substance.

While representing the cell, the half cell at which oxidation occurs is always written as placed to the left and the half cell at which reduction occurs is written on the right. The two parallel lines are drawn between –ve and +ve electrodes represents the interval connection of salt bridge.

Daniel cell: Daniel cell is an electrochemical cell constructed by coupling Zn and cu electrodes. In Daniel cell, Zn serves as –ve electrodes and Cu serves as +ve electrode.

Symbolically Daniel cell can be represented as

$$Zn_{(s)} \,/\, Zn^{2+}_{(aq)} \,//\, Cu^{2+}_{(aq)} \,/\, Cu_{(s)}$$

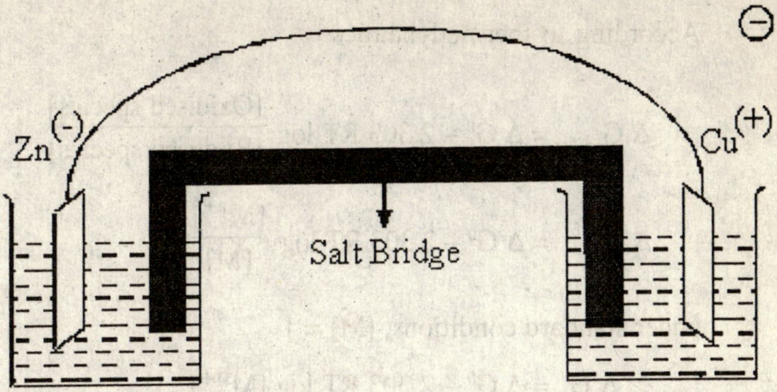

The electrode reaction are

at the anode $Zn \rightarrow Zn^{2+} + 2e^-$

at the cathode $2e^- + cu^{2+} \rightarrow Cu$

cell net reaction $Zn + cu^{2+} \rightarrow Zn^{2+} + Cu$

the emf of the cell is given by the equation

$$E_{cell} = E_r - E_r \text{ or } E_{cell} = E^{\oplus} - E_{(-)} \text{ or } E_{cell} = E_{cathode} - E_{anode}$$

Problems:

1. The standard electrode potentials of Zn and Ag electrode are – 0.76v and 0.80v. Construct the electrochemical cell using these two electrodes and represent the cell symbolically. Write the electrode, reaction and cell reactions and calculate the emf of the cell.

Symbolic representation of cell

$Zn / Zn^{2+} // Ag^+ / Ag$

and pressure is 1 atmosphere.

Derivation of Nerst equation.

Consider a metal M, in contact with its own ions. The electrode reactions is

$$M \rightarrow M^{n+} + ne^-$$

Where 'n' is the number of electrons involved in the reaction, either lost or gained.

According to thermodynamics.

$$\Delta G = \Delta G^0 - 2.303 \text{ RT} \log \frac{[\text{Oxidised species}]}{[\text{Reduced species}]}$$

$$\Delta G = \Delta G^0 - 2.303 \text{ RT} \log \frac{[M^{n+}]}{[M]}$$

under standard conditions, [M] = 1

$$\therefore \Delta G = \Delta G^0 - 2.303 \text{ RT} \log [M^{n+}]$$

we know that ΔG = -nEF and $\Delta G^0 = -nE^0F$

$$\therefore -\text{Nef} = -nE^0F - 2.303 \text{ RT} \log [M^{n+}]$$

Divide throughout by nF

$$E = E^0 + \frac{2.303 \text{ RT}}{nF} \log [M^{n+}]$$

$$E = E^0 + \frac{0.0591}{n} \log [M^{n+}]$$

Where, E is single electrode potential

E^0 is standard electrode potential

n is number of electrons involved in the reaction.

M^{n+} γ The concentration of the metal ion in the solution.

Electrode reactions.

At anode Zn γ $Zn^{2+} + 2e^-$

At cathode $2Ag^{2+}$ + 2e γ 2Ag.

Cell reaction Zn + $2Ag^{2+}$ γ 2Ag + Zn^{2+}

E_{cell} = E^0Ag – E^0zn

E_{cell} = 0.80 – (0.76)

E_{cell} = 1.56 v.

2. calculate the emf of the cell, given $E^0 sn^{2+} / sn^{+} = 0.15v$ and $E^0 Fe^{3+} / Fe^{2+} = 0.77v$.

$E_{cell} = E^0 / Fe^{2+} - E^0 sn^{2+}$

$E_{cell} = 0.77 - 0.15$

$E_{cell} = 0.62$ v.

Electrode Potential:

When a metal rod is in contact with the solution containing its own ions, there exist an electrical double layer at the interface of metal rod and its solution where there is an electrical double layer, there is a potential called electrode potential. It is denoted by a symbol 'E'.

Standard electrode potential: 'E^o'

It is the electrode potential measured when the metal ion concentration is 1mol/dm^3, temperature is 298k.

Reference Electrode:

Standard Hydrogen Electrode : SHE

Standard hydrogen electrode is a primary standard electrode with which we can measure, the potential developed in newly constructed other electrodes.

Construction: It consists of glass platinum foil to which a platinum wire is welded.

The platinum wire can be used for external connection. The platinum wire is kept in a glass tube. The glass tube is enclosed in an outer jacket. Outer jacket is having an inlet at the top and two holes at the bottom which are exactly at the middle of the platinum foil. The arrangement is dipped in exactly 1.0M HCl.

Working : Pure and dry H_2 gas is passed through the inlet at the top uniformly at one atmospheric pressure. Some of the H_2 molecule adsorbs on the surface of the platinum foil and the excess H_2 escapes out. There exist an equilibrium between hydrogen ions of the solution and H_2 molecules which are adsorbed on platinum foil.

The electrode reactions can be represented as

$$H_2 \underset{\text{Reduction reaction}}{\overset{\text{Oxidation reaction}}{\rightleftharpoons}} 2H^+ + 2e$$

Thus, set up electrode is called standard hydrogen electrode. The potential value of SHE is fixed as zero at all temperature.

Symbolically SHE can be represented as platinum H_2 / H^+

Use of SHE:

1. By making use of SHE, the potential of newly constructed electrodes can be measured.
2. With the help of SHE, it is possible to determine the P^H of a solution.

Limitations of SHE:

1. It is very difficult to prepare exactly 1.0M HCl
2. It is difficult to pass H_2 gas uniformly at 1 atmosphere pressure.
3. The H_2 gas must not contain even trace amount of impunities. Because impurities spoil the activity of platinum foil.
4. SHE cannot be used in presence of oxidizing (or) reducing agent.

Electrochemical series:

Electrode potential values of different single electrode are determined and when these electrodes are arranged in the increasing order of their reduction potential values to get a series. This series is known as electrochemical series.

In electrochemical series E^0 of standard hydrogen electrode is ± 0.0 and hence standard hydrogen electrode is placed in the middle of electrochemical series. The elements which have a tendency to undergo oxidation possess negative E^0 values. Which are placed in the upper portion of the electro chemical series. On the other hand, the elements which are readily undergoing reduction possess the positive E^0 values, which are placed below the standard hydrogen electrode in electrochemical series.

Electrode	Electrode reaction	E^0 values (V)
Li^+/Li	$Li^+ + e^- \gamma Li$	- 0.05 V
Zn^{2+}/Zn	$Zn^{2+} + 2e^- \gamma Zn$	- 0.76 V
Fe^{2+} / Fe	$Fe^{2+} + 2e^- \gamma Fe$	- 0.44 V
$2H^+ / H_2$	$2H^+ + 2e^- \gamma H_2$	0.0 V
Cu^{2+} / cu	$Cu^{2+} + 2e^- \gamma cu$	+ 0.34 V
Ag^+ / Ag	$Ag^+ + e^- \gamma Ag$	+ 0.80 V
Au^{3+} / Au	$Au^{3+} + 3e^- \gamma Au$	+ 1.50 V

The element which is kept at the top of electro chemical series is a strong reducing agent. An element in an electrochemical series, can reduce any other metal which is kept below in an electrochemical series.

Significance of electrochemical series :

1. The elements placed above the standard hydrogen electrode have tendency to denote electron. Hence these electrodes are good reducing agents.

2. The elements which are placed above the standard hydrogen electrode can displace H_2 gas from acids. In general an element with negative E^0 value displaces H_2 from acids.

3. The elements placed above the standard hydrogen electrode are electro positive and they undergo corrosion when exposed to air and moisture, but the metals which are below the standard hydrogen electrode are corrosion resistant.

4. An electrochemical cell of a required emf can be constructed by coupling two suitable electrodes. The suitable electrodes can be selected with the help of electrochemical series.

Reversible electrodes and cells :

A standard electrode or standard cell is one which gives constant and reproducible emf. There must not be change in emf with change in temperature. If there is any change in emf it should be very small which can be neglected. Standard hydrogen electrode is a primary reference electrode used to measure cell potential. It is very difficult to set up the electrode (SHE) in presence of impurity gases present in hydrogen. Therefore, secondary reference electrodes such as calomel electrode and Quin hydrone electrodes are used for finding the electrode potential.

Calomel electrode :

Calomel electrode is a reference electrode. It consists of a glass vessel having Hg at bottom. Hg is covered with a paste of mercurous chloride which is also called calomel. The vessel is completely filled with saturated solution of KCl. The potential of this electrode depends on the concentration of KCl solution. Electrical connections are made through a platinum wire dropped in the mercury at the bottom of the vessel. To connect with the other half cell, salt bridge is used. The electrode can be represented as KCl, Hg_2Cl_2 / Hg, pt.

The potential of the electrode for saturated KCl solution is given by 0.24N and for 1M KCl, 0.280V

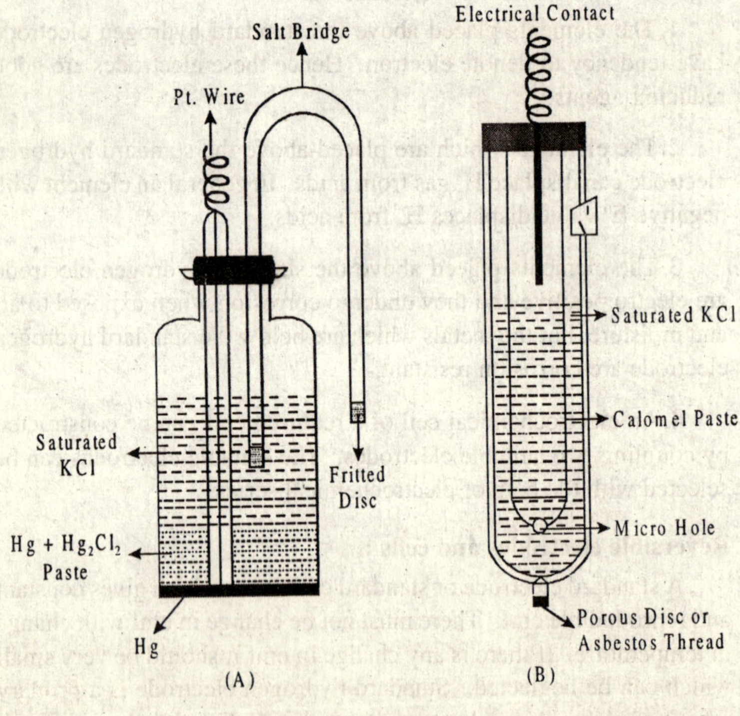

Calomel Electrode (a) with Salt Bridge (b) Compact Electrode.

Calomel electrode is simple to construct cell potential reproducible and does not vary with temperature.

Quinhydrone electrode:

Quinhydrone is an equimolar mixture of quinine and hydroquinone. When this matter is added into a solution containing H^+ ions, the following equilibrium exists.

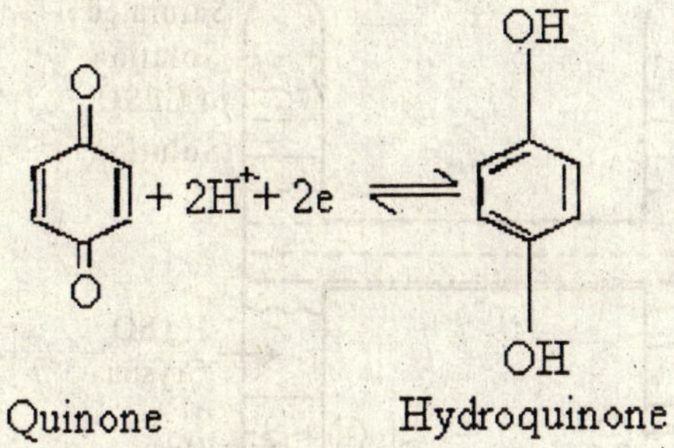

It can be also represented as

$$Q + 2H^+ + 2e \rightleftharpoons QH_2.$$

The platinum wire is dipped into the solution which can be used for external connection. Thus set up electrode is called Quinhydrone electrode.

E^0 of Quinhydrone = 0.699V. The electrode can be represented as Pt Q / QH_2 $H^{\oplus}$

Limitations :

1. It can not be used in presence of alkaline solution.
2. It can not be used in presence of any substance in the solution which can react with Q and QH_2.

Weston – cadmium cell :

It is a standard cell. It is used to measure the emf of other cells. Weston – cadmium cell gives a constant emf of 1.0186V.

Construction : Weston – cadmium cell consists of H – shaped glass vessel. Two Pt – wires are sealed at the bottom of the top limbs these Pt – wires can be used for external connection. One of the limb consists of Cd – amalgam above, where $CdSO_4$ crystals are placed. This serves as an anode. The other limb consists of Hg, over which Hg_2SO_4 crystals are placed. This serves as a cathode. The cell is completely filled with saturated solutions of $CdSO_4$. A little dilute H_2SO_4 is added to prevent hydrolysis. Symbolically Weston – cadmium can be represented as

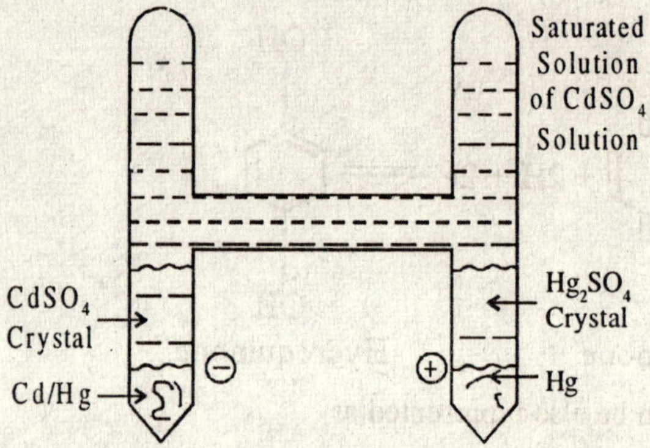

$Pt^{(-)}$ Cd / Cd^{2+} // Hg^+ / Hg, pt

At anode $\quad Cd \xrightarrow{\text{oxidation}} Cd^{2+} + 2e^-$

At cathode $\quad 2Hg^+ + 2e \longrightarrow 2Hg$

Cell reaction $\quad Cd + 2Hg^+ \rightarrow Cd^{2+} + 2Hg$

QUESTIONS CARRYING TWO MARKS

1) What are activity and activity coefficient ?
2) What is standard electrode potential?
3) What is an electrochemical series?
4) What are electrochemical cells?
5) What is solubility product?
6) What is the relationship between activity coefficient and ionic strength?
7) What is a common ion effect?
8) What is meant by conductance of a solution?

QUESTIONS CARRYING FOUR MARKS

9) Derive Nernst equation.
10) Explain the method of determining the pH of a given solution by using glass electrode.
11) Explain the principle and working of glass electrode.
12) What are ion selective electrodes? Explain its applications.
13) Explain the construction and working of calomel electrode.
14) How do you determine the solubility of sparingly soluble salt?

8

ACIDS, BASES AND BUFFERS

Modern Concepts of Acids and Bases

Arrhenius was the first person to explain acid-base concepts in aqueous medium. According to Arrhenius, an acid is a substance which on dissociation in aqueous medium to give H^+ ion and base is a substance which on dissociation to give OH^- ion in aqueous medium. But Arrhenius explained acid-base concept only in aqueous medium. Later Bronsted and Lowry explained acid base concept by considering exchange of only proton among the substances. According to lowry and bronsted concept.

Bronsted acids: CH_3COOH, C_6H_5OH, HCl

Bronsted bases: NH_3, H_2O

Bronsted-Lowry theory: An acid is a proton donor and a base is a proton acceptor. An acid-base reaction involves transfer of a proton from acid to a base.

$$H_2O + CH_3COOH \rightleftharpoons H_3^+O + CH_3COO^-$$

Conjugate acid base pairs: An acid formed by gaining of a proton by a base is called its conjugate acid. A base is formed by the loss of a proton by an acid is called its conjugate base. An acid and a base differ by a proton are called conjugate acid-base pairs.

Example: H_2O and OH^- and NH_3 and NH_4^+

This theory fails to explain the behavior of acidic oxides such as CO_2, SO_2, SO_3 etc, and basic oxides such as CaO, Na_2O, BaO etc.

Lewis concept: Lewis considered acid-base concept in terms of electron pairs. An acid is an electron acceptor which can accept a pair of electron. A Lewis acid contains an atom which is an electron deficient or has a vacant orbital in its valance shell.

Example : SO_2, CO_2, $AlCl_3$, BF_3, $FeCl_3$, SO_3 etc.

And base is an electron donor which can donate a pair of electron to other species. Lewis base contains an atom which is electron rich or contains one or more unshared pair of electrons in its valence shell.

Example : NH_3, H_2O, ROH, CaO, BaO etc.

Dissociation of Water

Water is the most abundant compound in nature. It is known that about $\frac{2}{3}^{rd}$ of earth's surface is occupied by water. It is considered to be a universal solvent as it dissolves many substances. Without water life would have been impossible on earth. Therefore, water is an important component amongst the other substances. In view of its wide applications as being a universal solvent, it becomes necessary to study its physical and chemical properties. Many other properties associated with water, one of the most important properties of it, is its dissociation. Since, it is a weak electrolyte, its ionization is limited. However, it undergoes ionization to a lesser extent leading to formation of H^+ and OH^- ions. Usually, H^+ exists as H_3O^+ (hydronium ion) in water. Hence its dissociation can be represented as

$$H_2O \rightleftharpoons OH^- + H^+$$

Using law of mass action, the dissociation of water can be written as

$$K_d = \frac{[OH^-][H^+]}{[H_2O]} \text{———————} (1)$$

K_d is known as dissociation constant of water.

$$K_d \times [H_2O] = [H^+][OH^-] \text{———————} (2)$$

Since the change in concentration of $[H_2O]$ is negligible, equation (2) can be written as

$$K_d = [H^+] [OH^-] \text{ ——————— (3)}$$

The product of $K_d \times [H_2O] = K_w$, Where K_w is called the **ionic product of water** and is defined as "the product of ions which are produced due to the dissociation of water" and is given by,

$$K_w = [H^+] [OH^-]$$

The product of concentration of $[H^+]$ and $[OH^-]$ ions formed during the dissociation of water has been confirmed that at 298K.

$[H^+] = 1 \times 10^{-7}$ and

$[OH^-] = 1 \times 10^{-7}$

$K_w = [1 \times 10^{-7}] [1 \times 10^{-7}]$

$K_w = 1 \times 10^{-14}$ and is constant at 298K

Hydrogen Ion Concentration - P^H

The famous scientist Oxeme Sorensen introduced a new scale in 1909 to decide the nature and strength of acids and bases. This new scale is called P^H scale. P^H is defined as "negative logarithm of molar concentration of hydrogen ions". The values of P^H ranges from 0 to 14

Mathematically P^H is given by

$$P^H = - \log [H^+]$$

P^H of a solution can also defined as negative logarithm of activity of hydrogen ion concentration and it can be written as

$$P^H = - \log [a_H^+] \text{ .... (1)}$$

Where a_{H^+} is the activity of H^+ ions and it is given by

$$a_H^+ = C_H^+ \times \gamma_{H^+}, \text{ ....(2)}$$

Where $C_H^+ \rightarrow$ The concentration of hydrogen ions

$\gamma_{H^+} \rightarrow$ The activity coefficient of hydrogen ions.

Since we deal with the dilute solutions, γ_{H^+} is almost equal to unity. Hence, $a_H^+ = C_H^+$

$$P^H = -\log C_H^+$$

Then, the P^H Can be defined as "the negative logarithm of molar concentration of hydrogen ions".

To Show That $P^H + P^{OH} = 14$

we know that, $Kw = [H^+]\,[OH^-]$

taking logarithm on both sides

$\log Kw = \log \{[H^+]\,[OH^-]\}$

$\log Kw = \log [H^+] + \log [OH^-]$

Changing the sign- $\log Kw = -\log [H^+] - \log [OH^-]$

At 298 K, $-\log 10^{-14} = P^H + P^{OH}$

$14 = P^H + P^{OH}$ or $P^H + P^{OH} = 14$

(1) For acidic solution: $P^H < 7$ or $P^{OH} > 7$

(2) For basic solution: $P^H > 7$ or $P^{OH} < 7$

(3) For neutral solution $P^H = 7$

Problems

(1) Calculate the P^{OH} of $\frac{1}{100}$ N sodium hydroxide

NaOH is strong mono acidic base

$[Base] = [OH^-] = 10^{-2}$

$P^{OH} = -\log [OH^-]$

$P^{OH} = -\log (10^{-2})$

$P^{OH} = 2$

$P^H + P^{OH} = 14$

$P^H + 2 = 14$

$P^H = 14-2$

$P^H = 12$

(2) Calculate the P^H of an aqueous solution 0.005M H_2SO_4

H_2SO_4 is dibasic acid

$[H^+]$ = 2 (acid)

$= 2 \times 10^{-3} \times 5 = 10^{-2}$

$P^H = - \log [H^+]$

$P^H = - \log (10^{-2})$

$P^H = 2$

(3) An aqueous solution contains 2×10^{-5} mole / dm^{-3} of KOH Calculate its P^H and P^{OH}.

[OH] = [base] = $[2 \times 10^{-5}]$

$P^{OH} = - \log [2 \times 10^{-5}]$

$P^{OH} = - [\log 2 - 5]$

$P^{OH} = 5 - \log 2$

$P^{OH} = 5 - 0.3010$

$P^{OH} = 4.6990$

$P^H = 14 - P^{OH}$

$P^H = 14 - 4.6990$

$P^H = 9.3010$

Determination of P^H of a Solution Using Standard Hydrogen Electrode and Calomel Electrode.

Principle:- Potential of hydrogen electrode depends on the $[H^+]$ and hence on the P^H of the solution. The hydrogen electrode is dipped in a solution whose P^H is to be determined. It is coupled with a standard electrode like calomel electrode to construct a cell. The e.m.f. of the cell can be measured by poggenderff's compensation method using Weston - Cadmium cell by knowing e.m.f of cell and E_{cell}, P^H of a solution can be calculated by the equation.

$$P^H = \frac{Ecell - Ecal}{0.0591}$$

Procedure:

The hydrogen electrode is dipped in a solution which is under examination. It is coupled with calomel electrode using salt bridge. The following electrochemical cell is constructed

$$Pt^-, H_2 \mid H^+ \parallel KCl, Hg_2 Cl_2 \mid Hg, Pt^+$$

The e.m.f. of the cell can be measured by using Potentiometer Bridge again by compensation method.

We know that, $E_{cell} = E\text{ cathode} - E\text{ anode}$

$$E_{cell} = E_{cal} - E_{SHE} \text{ .... (1)}$$

Consider the hydrogen electrode at which the hydrogen ions are in equilibrium with H_2 gas. The equilibrium can be represented as

$$H^+ + e^- \rightarrow 1/2\, H_2$$

By applying Nernst equation

$$E_{SHE} = E^0_{SHE} + \frac{2.303RT}{nF} \log \left[H^+\right]$$

$$E_{SHE} = 0 \text{ and } n = 1$$

$$E_{SHE} = 0.0591 \log (H^+)$$

$$E_{SHE} = -0.0591 [-\log (H^+)]$$

$$E_{SHE} = -0.0591.\, P^H$$

Substituting E_{SHE} in (1)

$$Emf_{Cell} = E_{cal} + 0.0591\, P^H$$

$$P^H = \frac{E_{cell} - E_{cal}}{0.0591}$$

Determination of P^H of a Solution by Using Quinhydrone Electrode

A Small quantity of quinhydrone powder is added to given solution whose P^H is to be determined. A platinum wire is dipped in it. It is coupled with a reference electrode such as calomel electrode and the following electrochemical cell is constructed.

$$Pt^{-}, Hg \mid Hg_2Cl_2.\ KCl \parallel QH^{\oplus} \mid QH_2\ Pt^{\oplus}$$

The emf of the cell is given by the equation

$$E_{cell} = E_{QH_2} - E_{cal} \quad ....(1)$$

Emf of the cell is determined by using Potentiometer Bridge by compensation method.

Consider the Quinhydrone electrode where the following equilibrium exists.

$$Q + 2H^{\oplus} + 2e^{-} \rightarrow ?QH_2$$

By applying Nernst equation

$$E_{QH_2} = E^0{}_{QH_2} + \frac{2.303}{nF} RT \log \frac{[\text{Oxidised species}]}{[\text{Reduced species}]}$$

$$E_{QH_2} = 0.699 + \frac{0.0591}{2} \log \frac{[Q][H^+]^2}{[QH_2]}$$

But $[Q] = [QH_2]$

$$E_{QH_2} = 0.699 + \frac{0.0591}{2} \log [H^+]^2$$

$$E_{QH_2} = 0.699 + 0.0591 \log [H^+]$$

$$E_{QH_2} = 0.699 - 0.0591 [- \log (H^+)]$$

$$E_{QH_2} = 0.699 - 0.0591\ pH$$

By substituting E_{QH_2} in (1)

$$E_{cell} = 0.699 - 0.591\ P^H - E_{cal}$$

$$0.0591\ P^H = 0.699 - E_{cell} - E_{cal}$$

$$P^H = \frac{0.699 - (E_{cell} + E_{cal})}{0.0591}$$

Effect of salt on the dissociation of weak acids and bases:

Salt on dissociation gives anion and a cation in aqueous medium. The cation and anion of a salt reacts with water to give acidic or alkaline solution.

Consider sodium acetate which is a salt of weak acid and strong base. When dissolved in water it undergoes completely dissociation to produce acetate ion and sodium ion as follows:

$$CH_3\,COO\,Na \rightleftharpoons CH_3\,COO^- + Na^{\oplus}$$

The anion of the salt interacts with water to give a weak electrolyte acetic acid.

$$CH_3\,COO^- + H_2O \rightleftharpoons CH_3\,COOH + OH^{\ominus}$$

Due to this reaction, the concentration of H^+ ion in the solution decreases. Since anion interacts with water it is called anionic hydrolysis. The solution contains more of OH^- ions. Hence aqueous solution of sodium acetate is alkaline to litmus. By applying law of mass action, the hydrolysis constant is given by

$$K_h = \frac{[CH_3\,COOH]\,[OH^-]}{[CH_3\,COO^-]\,[H_2O]}$$

Concentration of water is taken as unity

$$K_h = \frac{[CH_3\,COOH]\,[OH^-]}{[CH_3\,COO^-]}$$

$$K_h = \frac{[CH_3\,COOH]\,[OH^-]\,[H^+]}{[CH_3\,COO^-]\,[H^+]}$$

$$K_h = \frac{K_w}{K_a} \quad OR \quad K_a = \frac{K_w}{K_h}$$

Where K_a – Dissociation constant of weak acid.

K_h – Hydrolysis constant.

K_w – Ionic product of water.

Consider the salt NH_4Cl, which is a salt of strong acid and weak base. When dissolved in water it undergoes complete dissociation giving NH_4^+ and Cl^-.

$$NH_4Cl \rightarrow NH_4^+ + Cl^-.$$

The cation of the salt reacts with water forming a weak electrolyte NH_4OH.

$$NH_4 + H_2O \rightarrow NH_4OH + H^+.$$

Due to this reaction, the concentration of H^+ ions in the solution increases when compared to $[OH^-]$. Hence aqueous solution of NH_4Cl is acidic in nature.

By applying law of mass action, hydrolysis constant is given by

$$K_h = \frac{[NH_4OH]\,[H^+]}{[NH_4^+]\,[H_2O]}$$

The concentration of water is taken as unity.

$$K_h = \frac{[NH_4OH]\,[H^+]}{[NH_4^+]}$$

$$K_h = \frac{[NH_4OH]\,[H^+]\,[OH^-]}{[NH_4^+]\,[OH^-]}$$

$$K_h = \frac{K_w}{K_b}$$

OR $$K_b = \frac{K_w}{K_h}$$

Where K_b – Dissociation constant of weak base

K_h – Hydrolysis constant.

K_w – Ionic product of water.

Consider a salt of CH_3COONH_4, which is a salt of weak acid and weak base. When dissolved in water, it dissociates completely as follows.

$$CH_3COONH_4 \rightleftharpoons CH_3COO^- + NH_4^+.$$

Both anion and cations of salt undergoes hydrolysis to give weak electrolyte, acetic acid and ammonium hydroxide.

$$CH_3COO^- + NH_4^+ + H_2O \rightarrow CH_3COOH + NH_4OH.$$

By applying law of mass action the hydrolysis constant is given by

$$K_h = \frac{[CH_3COOH]\,[NH_4OH]}{[CH_3COO^-]\,[NH_4^+]\,[H_2O]}$$

$$[H_2O] = 1, \quad K_h = \frac{[CH_3COOH]\,[NH_4OH]}{[CH_3COO^-]\,[NH_4^+]}$$

$$K_h = \frac{[CH_3COOH]\,[NH_4OH]\,[H^+]\,[OH^-]}{[CH_3COO^-]\,[NH_4^+]\,[H^+]\,[OH^-]}$$

$$K_h = \frac{K_w}{K_a\,K_b}$$

$$\text{or} \qquad K_a = \frac{K_w}{K_h\,K_h}$$

Where

K_a & K_b – Dissociation constant of weak acid and weak base respectively.

K_h – Hydrolysis constant.

K_w – Ionic product of water.

Buffer Solution: Buffer is an aqueous system, that can resist change in their P^H. Buffers are a mixture of a weak acid and its conjugate base or a weak base and its conjugate acid. The change in P^H of the solution on addition of an acid or base is controlled by using buffer solutions.

Example: Mixture of acetic acid and sodium acetate. Mixture of ammonium hydroxide and ammonium chloride.

Acidic Buffers

A mixture of acetic acid and sodium acetate is an acidic buffer. Addition of a small quantity of an acid will not change the P^H of the buffer. The H^+ ions from the acid are neutralized by the acetate ions.

$$CH_3\,COONa \rightarrow CH_3\,COO^- + Na^+_{(aq)}$$

When an acid like HCl is added

$$H^+ + CH_3\,COO^- \rightarrow CH_3\,COOH$$

When a base like NaOH is added, OH^- ions from the base are neutralized by acetic acid.

$$OH^- + CH_3\,COOH \rightarrow CH_3\,C\,OO^- + H_2O$$

Thus addition of acid and base does not alter the P^H of the buffer.

Basic Buffer: A Mixture of ammonia and ammonium chloride is a basic buffer. On addition of a small quantity of an acid, H^+ ions from acid are neutralized by NH_3

$$H^+ + NH_3 \rightarrow NH_4^+$$

When a base like NaOH is added, the OH^- ions from base are neutralized by NH_4^+ ions

$$OH^- + NH_4^+ \rightarrow NH_3 + H_2O$$

Thus addition of small quantity of an acid and a base will not alter the P^H of the buffer.

P^H is defined as negative logarithm of H^+ ion concentration. P^H of a buffer solution depends on the P^{ka} value of weak acid and ratio of salt to acid concentration. The relation between P^H, P^{ka} and the ratio of concentration of ionized and unionized form of an acid or a base is given by Henderson – Hasselbalch equation.

Consider the dissociation of weak acid HA

$$HA \rightarrow H^{+} + A^{-}.$$

Apply law of mass action, $K_a = \frac{[H^+][A^-]}{[HA]}$

K_a is called dissociation constant of a weak acid

$$K_a [HA] = [H^+][A^-]$$

$$[H^+] = K_a \frac{[HA]}{[A^-]} \text{ - - - - - - - - x}$$

Take logarithm on both side of the equation x.

$$\ln [H^+] = \ln K_a \frac{[HA]}{[A^-]}$$

$$\text{In } [H^+] = \ln K_a + \ln \frac{[HA]}{[A^-]}$$

$$\text{-In } [H^+] = -\ln K_a - \ln \frac{[HA]}{[A^-]}$$

$$P^H = P^{Ka} + \ln \frac{[A^-]}{[HA]}$$

Note: The above equation is called Henderson – Hasselbalch equation

If we know the dissociation constant of the weak acid K_a, P^H can be calculated .

$$P^H = P_{K_a} + \log \frac{[salt]}{[acid]}$$

If we know the dissociation constant of the base K_b, P^{OH} can be calculated using the formulae

$P^{OH} = -\log [OH^-]$ $P_{Kb} = \log K_b$

$$P^{0H} = P_{K_b} + \log \frac{[salt]}{[base]}$$

Buffer capacity can be defined as , the number of moles of an acid or base to one liter of the buffer solution, so as to change its P^H value by one unit. In a buffer solution $\frac{[salt]}{[acid]}$ or $\frac{[salt]}{[base]}$ shoud not be more than that of 10 or less than1/10. If it exceeds these limits, the capacity decreases. P^H range of buffer = $P_{K_a} \pm 1$ or $P_{K_b} \pm 1$

All biological functions are P^H dependent. The activity of biological macro-molecules and catalytic activity of enzymes are affected by P^H. Enzymes show maximum activity at a particular P^H called optimum P^H. The major buffering system found in cellular fluids involve phosphate buffer, bi-carbonate buffer etc. The P^H of the blood plasma and tears is measured to 7.4, which are slightly alkaline. If the P^H of the blood plasma exceeds 7.4, that condition is called alkalosis which leads to hysteria, ankle joints and weakening of wrist. If the P^H of blood plasma is lower than 7.4, condition is called acidosis, which leads to pneumonia & finally comoa state.

Compounds containing both amino ($-NH_2$) group and carboxylic group (-COOH) are called amino acids. Those are building blocks of proteins. All naturally occuring proteins on hydrolysis yields 20 different types of amino acids. Since amino acids contain both amino (basic) group and carboxylic group, they behave both as acids and bases. Hence these are called amphoteric compounds. The structure of amino acids in molecular form can be written as

```
R – CH – COOH
     |
    NH2.
```

On the basis of the above structure, it is not possible to explain solubility in H_2O, non – volatility, crystalline in nature and high melting point and boiling point possessed by amino acids. Therefore, in order to explain these properties conveniently, the amino acids are represented as a dipolar ion or a zwitter ionic from equation

$$\begin{array}{l} R-CH-N^{+}H_3 \\ \quad\;\; | \\ \quad\;\; COO^{-} \end{array}$$

Zwitter ion

When the P^H of a solution containing amino acids is decreased by adding acid ($P^H < 7$), the amino acid is converted into cation. under the influence of electric fields, cation move towards cathode The solution containing amino acid is increased by adding base ($P^H >$ 7), the amino acid is converted into anion. Under the influence of electric field, it will move towards anode. At particular P^H, the amino acid neither move towards cathode nor towards anode. At this P^H, the amino acids exist as zwitter ion. Such a P^H is called **isoelectric P^H** or **Iso–electric point** of an amino acid.

Titration of glycine with an alkali and to indicate P^{Ka} and P^I (isolectric point):-

The titration curve for the titration of glycine with alkali as shown in the graph.

A) $P^H = P^k$, the α-carboxyl group of cation and zwitter ion exists in equal concentration. At $P^H = P^{k2}$, amino group of amino acid and zwitter ions exists in equal concentration. At $P^H = P^I$, amino acids mainly exists as zwitter ion.

$$\text{The Isoeletric point, PI} = \frac{P^{K_1} + P^{K_2}}{2}$$

For glycine $P^{K1} = 2.4$ and $P^{K2} = 9.8$

$$\therefore \text{PI} = \frac{2.4 + 9.8}{2} = \frac{12.2}{2} = 6.1$$

Cells and organisms maintain a specific and constant P^H. Constant P^H in biological system achieved primarily by biological buffers. Biological control of P^H of cells and body fluids is an important aspect of metabolism and cellular activities.

1. Bicarbonate – Carbonate buffer system.

Carbonate – bicarbonate buffer system is the main buffer in blood plasma. It is a mixture of carbonic acid bicarbonate and mixtures.

$$H_2CO_3 \rightarrow H^+ + HCO_3^-$$

At equilibrium condition.

$$K = \frac{[H^+][HCO_3^-]}{[H_2CO_3]}$$

The concentration of H_2CO_3 in body fluid depends on the concentration of CO_2, dissolved in water which is obtained by tissue metabolism when the ratio of $[HCO_3^-] / [H_2CO_3]$ becomes 20, the P^H of blood plasma maintained at 7.4. Inside the cell, the concentration of HCO_3^- is less and hence no importance.

2) Phosphate buffer system:

Phosphate buffer works in the extra cellular fluid at P^H ranges from 6.4 to 7.4. It is a combination of $H_2PO_4^-$ and HPO_4^{2-}.

The concentration of phosphate buffer in the extra cellular body fluid will be equal to 8% of bicarbonate buffer present in the body fluid. The phosphate buffers works effectively at P^H range between 6.0 and 6.9 in intracellular fluid.

$$H_2PO_4^- \rightarrow HPO_4^{2-} + H^+$$

In red blood cells, hemoglobin buffer plays an important role. Oxygenation and deoxygenation of hemoglobin influences the buffering capacity of hemoglobin. Carbon dioxide which is produced during tissue metabolism, combines with H_2O to form carbonic acid. Oxy hemoglobin (HbO_2) on losing oxygen to form de-oxy hemoglobin [Hb^-]. Deoxyhemoglobin readily accepts H^+ from H_2CO_3 to HCO_3^- & HHb.

$$H_2O + CO_2 \rightarrow H_2CO_3$$
$$HbO_2 \rightarrow Hb^- + O_2$$
$$Hb^- + H_2CO_3 \rightarrow HHb + HCO_3^-$$
$$H^+ + HCO_3^- \rightarrow H_2CO_3.$$

Determination of P^H Value of Buffer by Colour Comparison Method:

A buffer solution is used to maintain constant P^H of the reaction mixture. A mixture of weak acid and its salt or weak base and its salt is known as buffer solution. We can determine approximate P^H of unknown solution, by comparing the intensity of the colour developed in the unknown solution with the colour of the standard buffer solutions. Using the universal indicator, exact P^H of the unknown solution is determined by preparing the buffer solutions with in the approximate range of P^H The mixture of acetic acid and sodium acetate is an example of acidic buffer and ammonium hydroxide and ammonium chloride is an example of basic buffer.

To find the approximate P^H of unknown solution, prepare a series of standard buffer solution of P^H ranging from 1 to 14. Take each 10ml of standard buffer solutions in different test tubes and number the test tube from 1 to 14. Add two drops of universal indicator into each test tube. The indicator develops different intensity of colour in different buffer media.

Take the unknown solution, whose P^H is to be determined in another test tube and add two drops of indicator. Compare the intensity of colour developed in unknown solution, with the colour of the standard buffer solutions. Find out approximate range of P^H of the unknown solution.

Let the approximate P^H of the unknown solution is 6: To find the exact P^H of the unknown solution, again prepare a series of buffer solutions ranging from 6.0 to 7.0 with 0.1 unit difference in P^H values.

Again take 10 ml of each buffer solutions in different test tubes and add 2 drop of indicator into different test tubes. The indicator develops different intensity of colour with the different standard buffer solutions.

Now again compare the intensity of colour of unknown solution, with standard buffer solutions and there by find the exact P^H of unknown solution.

Note: [During the determination of approximate P^H of the unknown solution, find out the solution, whether it is acidic or basic and choose appropriate indicator]

Limitation of Colorimetric method:

Not suitable for very accurate estimation.

Use of Buffer Solutions:

1) Buffer solutions are used to maintain constant P^H in analytical chemistry experiments.
2) Buffer solutions of phosphates and ammonium salts are used in soil and agricultural chemistry
3) The P^H of the blood is controlled by the buffering agent serum proteins. If the P^H of human blood gets beyond the range of 7.5 - 7.6, it results in coma stage. P^H of the blood is also controlled by HCO^-_3 -H_2CO_3 buffer.
4) All biological activity is P^H – dependent, all enzymatic reactions in the body are controlled by P^H

Indicator :

Indicators is a substance which helps in carrying out titrations between acids and bases. These substances give particular colours in acidic and basic media. Hence, a sudden change in the colour is produced in the solution at the neutralization point. Phenolphthalein and methyl orange are examples of acid – base indicators. Redox indicators are organic compounds which have different colours in the oxidized and reduced forms. Diphenyl amine, methylene blue, diphenyl benzidine and diphenyl sulphonic acid are examples for redox indicators.

According to Oswald, an indicator is an organic weak acid or weak base, on ionization causes change in its colour. An acidic indicator such as phenolphthalein must possess a coloured anion while a basic indictor such as methyl orange must possess coloured cations. Unionized molecules of the indicator possess different colour than the ions.

Universal indicators show a colour change over a large p^H range. A universal indicator is a mixture of several indicators with a wide range of p^H response. It behaves like a single indicator. A mixture of suitable amounts of alcoholic solution of phenolphthalein, methyl orange, methyl yellow, bromothymol blue and thymol blue in sodium hydroxide solution is yellow in colour and this solution acts as a universal indicator showing colour changes at different p^H values.

Colour	Red	Orange	Yellow	Green	Blue
P^H	2	4	6	8	10

Phenolphthalein is a weak organic acid, exist mostly unionized form. These unionized molecule are colour less.

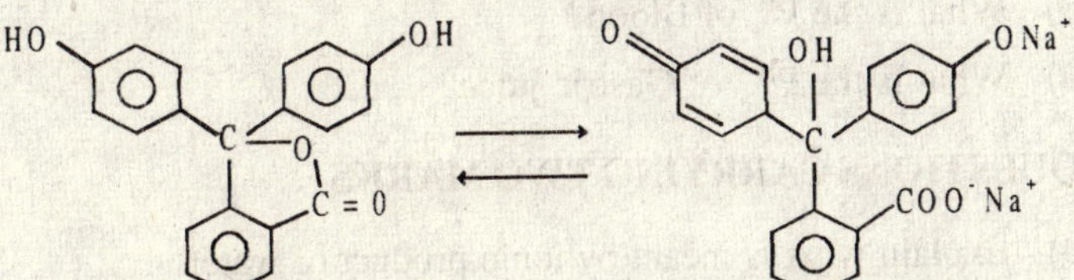

Phenolphthalein (colourless) (Pink colour)

Addition of acid increases H^+ ion concentration in the solution and the equilibrium shifts towards left. The solution becomes colourless. On addition of base into the solution containing phenolphthalein, OH^- of the base reacts with H^+ to form water. The equilibrium shifts towards right. The solution turns pink due to the formation of more anionic phenolphthalein.

Methyl orange is a weak organic base. This also works as indicator similar to phenolphthalein.

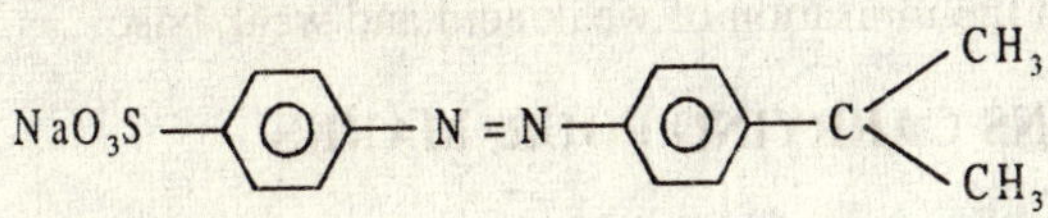

Methyl orange

Methyl orange $\rightleftharpoons$ Cationic form

Yellow Orange Red

QUESTIONS CARRYING ONE MARK

1) Define an acid? Give an example.
2) Define a base? Give an example.
3) Arrange the following in order of increasing ionic strength. K HF, HCl, HBr, HI.
4) Which of the following bases is the weakest and why. Cl^-, I^-, B_n^-, F^-.
5) Define the term P^H.
6) What is a buffer solution?
7) Give any two uses of buffer solution.
8) What is the P^H of Blood?
9) What is the P^H of Gastric juice?

QUESTIONS CARRYING TWO MARKS

10) Explain what is meant by ionic product of water.
11) Show that $P^H + P^{OH} = 14$
12) Explain the Buffer action of acidic buffer
13) Explain the Buffer action of a basic buffer
14) Describe the buffer action of ammonia buffer.
15) Describe the effects of salts on dissociation of acids
16) Explain the buffer action of Carbonate – bi carbonate in blood.
17) How is P^H of Red blood cells maintained.
18) Explain lewis acid - Base theory.
19) Explain the ionisation of weak acid and weak base.

QUESTIONS CARRYING FOUR MARKS

20. How is P^H of a solution is determined by using hydrogen and calomel electrode?
21. Calculate the P^H of a solution containing 0.02 MHCl. and 0.02 M NaOH

22. If a solution contains 6×10^{-5} mole / dm^{-3} of NaOH, calculate its P^H and P^{OH}

23. How is the P^H of a solution determined using Quinhydrone method.

24. Explain the determination of P^H by colour comparison method.

9
ADSORPTION

The surface of a solid or liquid is different from that in the interior. The molecule which is present in the interior of a liquid is completely surrounded by other molecules on all side and therefore intermolecular forces of attraction are exactly equal in all directions. However, a molecule at the surface of a liquid is surrounded by a large number of molecules in liquid phase and some molecules in the vapor phase. Due to this the molecules lying on the surface experience net inward force of attraction which is the cause for surface tension. Similar inward force of attraction exist at the surface of the solids. Because of these forces of attraction existing on the surface of solids and liquids, they have the property of retaining other molecules on their surface if they come in contact. This phenomenon of attracting and retaining the molecules or a substance on the surface of a liquid or a solid resulting into a higher concentration of the molecules on the surface is called adsorption.

The substance which adsorbs is known as adsorbent and the substances which gets adsorbed are called adsorbates.

Example:

Charcoal - Adsorbent

Colour matter - Adsorbate

Charcoal adsorb colouring matter. Therefore it is used in decolourisation process.

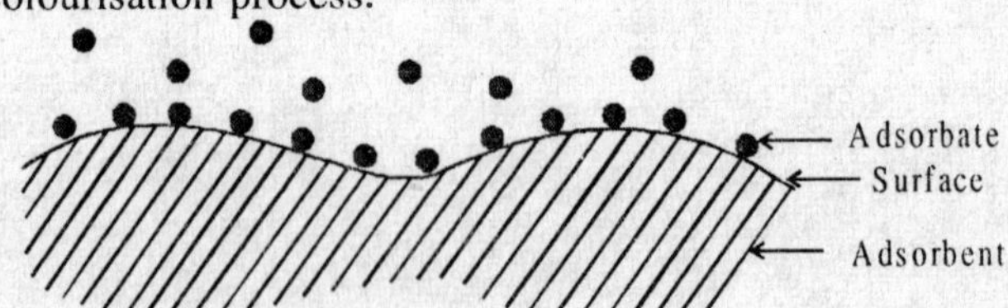

Fig. 9.1 Adsorption : Concentration at the surface

Types of Adsorption

Depending on the type of attractive forces between adsorbent and adsorbate, the adsorption is classified into two types these are,

i) Physical adsorption or Vander waals adsorption

ii) Chemical adsorption or activated adsorption.

Physical Adsorption

If a gas is held (adsorbed) on the surface of a solids by Vander waals forces of attraction without resulting into the formation of any chemical bond between the adsorbate and adsorbent, then it is called physical adsorption.

The heat of adsorption for physical adsorption is low (about 20-40 kJ mol^{-1})

Example: Adsorption of chlorine on the surface of charcoal

Chemical Adsorption

When a gas is held on to the surface of a solid by forces similar to those of a chemical bond. The type of adsorption is called chemisorption or chemical adsorption. The heat of adsorption is high compare to physical adsorption. (about 40 - 400 kJ mol^{-1}).

Example: Adsorption of gases like hydrogen on to metal surfaces like Nickel and platinum.

Differences between physical adsorption and chemical adsorption

Physical adsorption	*Chemical adsorption*
1. Weak Vander waal's forces are responsible for adsorption.	1. Strong Valence forces are responsible for adsorption.
2. The interaction between adsorbent and adsorbate is nonspecific.	2. The interaction between adsorbent and adsorbate is specific.
3. It is a reversible in nature	3. It is irreversible process.
4. Heat of adsorption is low (20 – 40 kJ mol^{-1})	4. Heat of adsorption is high (about 400 kJ mol^{-1})
5. Multilayer of adsorbates can be formed.	5. Generally monolayer is formed.
6. Decreases with increase in temperature.	6. First increases and then decreases with increase in temperature.

Adsorption of Gases by Solids (Charcoal, Silica, Alumina)

It is observed that the surface of solid materials contain residual forces, it is these residual forces responsible for retaining the molecules of other species. As the adsorbate molecules remain only on the surface and do not penetrate into bulk of the solid, the concentration is very high on the surface.

Examples of solids which are adsorbing: charcoal, silica, alumina etc.

The extent of adsorption depends on the surface area of the solids. The larger the surface area of the adsorbing agent, the greater is the adsorption, therefore, matter in the finely divided state or colloidal state exhibit great powers of adsorption. Therefore, finely divided metals like Nickel, platinum, and porous substances like charcoal fuller's earth, silica gel etc provides large surface area and are well known solid adsorbents. Thus the molecules adsorbed on the surface, show a distinct pattern. They are oriented and arranged in a definite manner relative to the adsorbing surface as well as to themselves.

Determination of Adsorption of Gases on to Solid Surface

Many methods have been used for the determination of adsorption of gases on solid adsorbents. In one method, the determination of extent of adsorption is as follows. The gas is taken in a container of known volume at given temperature. The pressure of the gas is measured on a monometer attached to the vessel. The adsorbent is then introduced into the vessel by a suitable device. Adsorption takes place fairly fast and the pressure of the gas falls. The decrease in the pressure of a gas is noted on a monometer. By knowing the fall in pressure, the quantity of gas adsorbed can be calculated.

Factors Affecting the Adsorption of Gases by Solids

Most of the solids have capacity to adsorb the gases to some extent. However, the extent of adsorption of gases by solids depends on many factors some of the factors are mentioned below.

i) *Nature and surface area of the adsorbent:* The same gas may have different degree of adsorption with different solids at the same temperature.

Further, it is observed that greater the surface area of the adsorbent, greater is the volume of the gas adsorbed. Charcoal and silica gel are good adsorbents because they have highly porous structure and hence more surface area.

ii) *Nature of the gas being adsorbed:* Different gases are adsorbed to different extents by the same adsorbent at the same temperature.

iii) *Temperature:* It is observed that the adsorption decreases with increase in temperature. Like any other equilibrium, adsorption process involves equilibrium between adsorption and desorption. On increasing temperature desorption (evoparation) of gas molecules will take place from the surface of solids. Since adsorption is also exothermic process, high temperature decreases the adsorption process.

iv) *Pressure:* At constant temperature, the adsorption of gas increases with increase in pressure. It is found that at low temperatures, the adsorption of a gas increases very rapidly as the pressure is increased.

Heat of adsorption

The amount of heat evolved when one mole of gas or vapour is adsorbed on a solid is known as molar heat of adsorption. This value depends on the nature of the gas being adsorbed. Generally adsorption is accompanied by decrease in enthalpy of the system. There ΔH value is invariably negative. The adsorption is also accompanied by decrease in entropy of the system. i.e., $T\Delta S$ of the process is also negative. The adsorption is thus accompanied by decrease in enthalpy as well as decrease in entropy of the system.

The free energy charge, ΔG, for any feasible process, as given by the well known thermodynamic equation.

$$\Delta G = \Delta H - T\Delta S.$$

As adsorption proceeds, the heat of adsorption per mole of the adsorbate goes on decreasing i.e., ΔH becomes less negative. Finally ΔH becomes equal to $T\Delta S$ and ΔG becomes Zero. This is the stage at which equilibrium is reached.

Adsorption isotherms

Adsorption isotherms are graphical representations of relationship between the rate of adsorption and pressure. At constant temperature, there exist some definite relationships between the amount of gas adsorbed and the equilibrium pressure. Such relationships are known as adsorption isotherms.

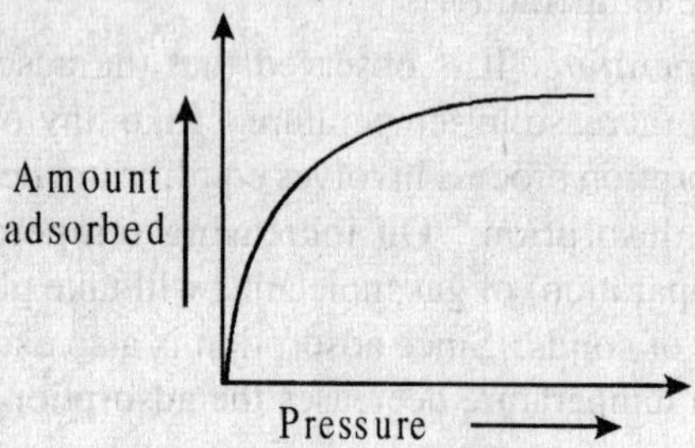

The adsorption isotherm is first explained by Freundlich and later improved by Langmuir. Now we shall derive the equations for Freundlich and Langmuir adsorption isotherms.

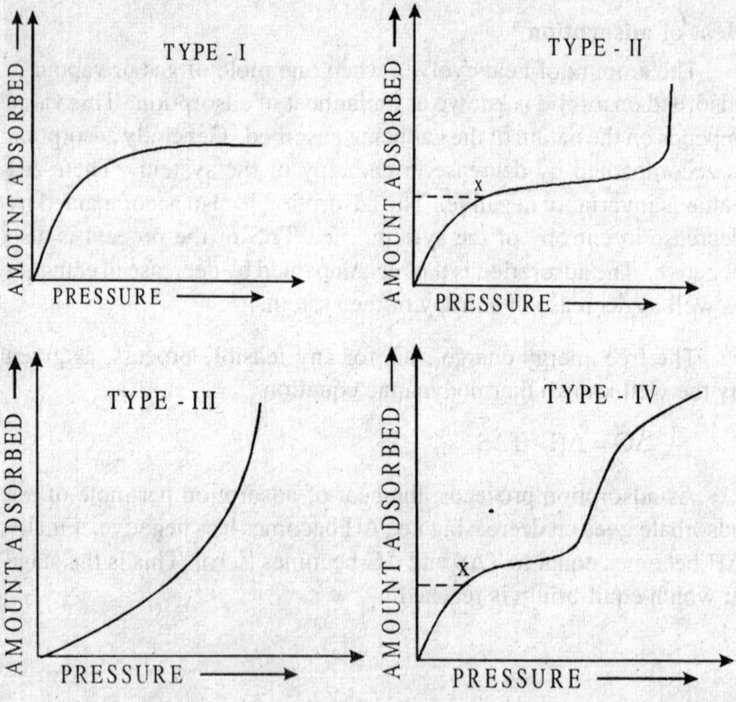

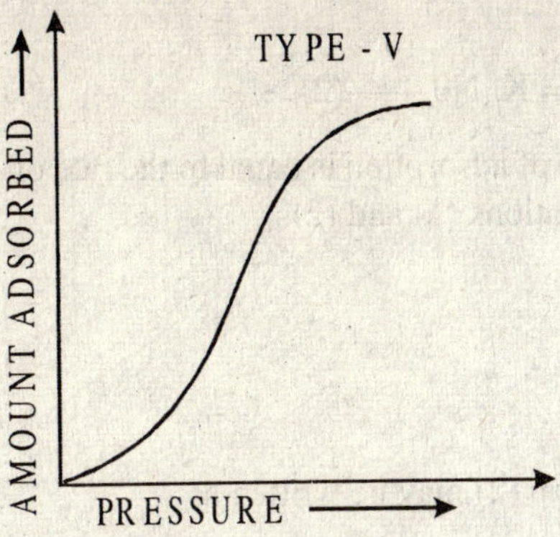

Types of adsorption isotherms

Theory of Adsorption

Langmuir Adsorption Isotherm

In 1916 Irving Langmuir proposed his theory of adsorption of a gas on the surface of a solid. He considered the surface of the solid to be made up of elementary sites each of which could adsorb one gas molecule. It is further assumed that a dynamic equilibrium exists between the adsorbed molecules and the free molecules.

If A is the gas molecule and M is the surface site, then

$$A\ (g) + M\ (surface) \underset{K_d}{\overset{K_a}{\rightleftharpoons}} AM$$

Where Ka and Kb are the rate constants for adsorption and desorption respectively. Now, the rate of adsorption is proportional to the pressure of 'A', viz., P_A and the number of vacant sites on the surface, viz., N (1 - θ). Where N is the total number of sites and θ is the fraction of surface sites occupied by the gas molecules i.e.,

$$\theta = \frac{\text{Number of adsorption sites occupied}}{\text{Number of adsorption sites available}}$$

Thus, the rate of adsorption = Ka P_A N(1 – θ) ——(1)

The rate of desorption is proportional to the number of adsorbed molecules, Nθ.

Thus, the rate of desorption = K_b Nθ ——(2)

Since at equilibrium, the rate of adsorption is equal to the rate of desorption, we can write the equations (1) and (2)

$Ka\, P_A\, N\,(1 - \theta) = K_d\, N\theta$

Or $KP_A\,(1 - \theta) = 0$

Where $K = \dfrac{K_a}{K_b}$ then equation (2) may be written as

$$\frac{1-\theta}{\theta} = \frac{1}{KP_A}$$

$$\frac{1}{\theta} - 1 = \frac{1}{KP_A}$$

$$\frac{1}{\theta} = \frac{1}{KP_A} + 1 = \frac{1+KP_A}{KP_A}$$

$$\theta = \frac{KP_A}{1+KP_A} \text{ ------ (3)}$$

The equation (3) is called Langmuir adsorption isotherm.

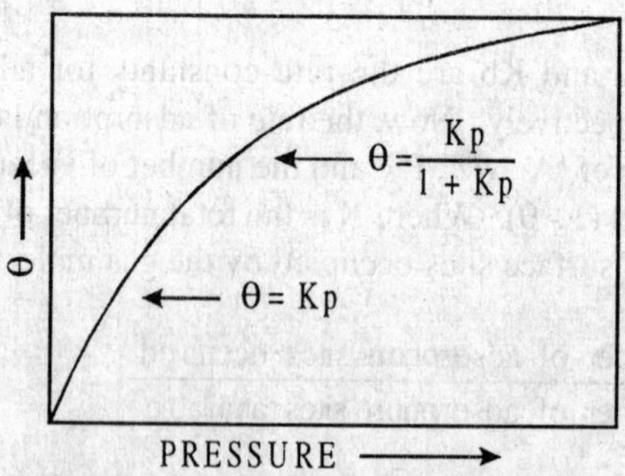

Fig. 9.2: The Langmuir plot, Kinetics of the gaseous reaction on solid surface.

Freundlich Adsorption isotherm

Since the adsorption is invariably accompanied by evolution of heat, therefore, in accordance with Le – Chatelier's principle, the magnitude of adsorption depends on temperature and pressure. Thus, decrease of temperature and increase of pressure both tend to cause increase in the magnitude of adsorption of a gas on a solid. The adsorption isotherm showing the magnitude of adsorption of a gas on a solid. The adsorption isotherm showing the relationship between the magnitude of the adsorption and the pressure is known as Freundlich adsorption isotherm. This is mathematically expressed by the equation.

$$A = KP^n \text{ ——— } (1)$$

Where 'A' is the amount of gas adsorbed per unit mass of the adsorbent at a pressure P and K and n are constants depending upon the nature of the gas and the nature of the adsorbent.

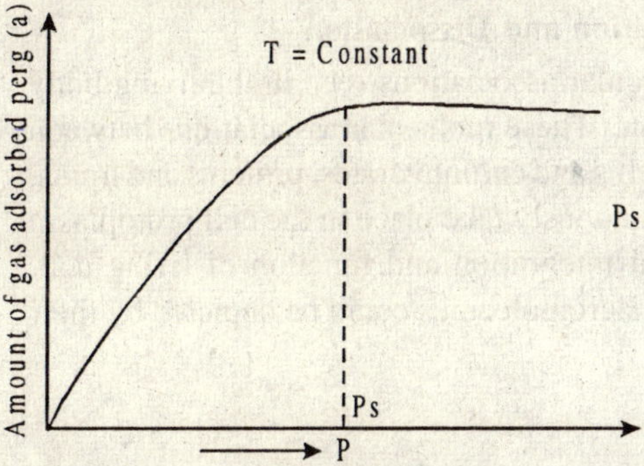

Fig : Frcundlich adsorption isotherm

Fig. 9.3: Freundlich adsorption isotherm.

Applications of Adsorption

Adsorption is associated with many of the reactions taking place in the living body. It also has various laboratory applications, industrial applications as well as various technical applications.

1. Catalysis

The velocity of any chemical reaction depends to a great extent on the active masses of the reactants, provided other factors are in optimum conditions. Active mass and concentration are synonymous and an increase in concentration of the components of the reaction facilitates the reaction. In the reactions taking place in the protoplasm, adsorption plays a very important role. The enzymes being colloidal in nature, have large surfaces and surface adsorption of reactants is a requisite for the enzymatic reaction to proceed smoothly. The colloidal molecules are interspersed with a network of membranes which possess enormous surfaces. Adsorption taking place at these membranes promotes several vital chemicals.

2. *Action of Drugs and Poisions*

Substances capable of reducing interfacial tension tend to accumulate at the interfaces. To this class of surface active agents belong various drugs and poisons. These substances, as a consequence of reducing interfacial tension, are concentrated at the interfaces and exert their effect from that location. Selective adsorption of the drugs may thus be responsible for specific action.

Macro molecular Association and Dissociation

Various types of molecular associations seen in the living body are as a result of adsorption. These molecular associations between proteins and proteins, proteins and carbohydrates, proteins and lipids, and proteins and salts continuously take place in the cell protoplasm and are vital for structural integration and function of living cell. Similarly dissociation of macromolecules could be depicted by their adsorption behaviours.

Histological studies

Adsorption need not be followed by absorption, asorption depends upon the concentration of the molecules at the surface or interface. Acidic and basic dyes which can be distinguished by their staining power are important tools not only in histological studies, but also in experimental physiology.

Adsorption in safety devices

A bacterial toxin, harmone, or a mineral poison may be inactivated by adsorption. Ferric hydroxide, a good adsorbent is often used as an antidote in cases of poisoning by arsenic. Gas masks are devices containing an adsorbent or a series of adsorbents which purify the air for breathing by adsorbing the poisonous gases and vapours from the atmosphere.

Chromatography

Important laboratory applications of adsorption is the recovery and concentration of vitamins, proteins, and other biological substances by the method of chromatographic analysis.

QUESTIONS CARRYING TWO MARKS

1. What is adsorption? How does it differ from absorption?
2. How is adsorption affected by (i) increase in surface area (ii) increase in pressure.
3. What is an adsorption isotherm? Give one example
4. Define the term adsorbent and adsorbate.
5. What is physical adsorption? Give an example
6. What is chemical adsorption? Give an example
7. What is heat of adsorption?
8. Write the equation which is used to calculate heat of adsorption?
9. Why is gas adsorbed on the surface of liquid or a solid?
10. Write the expression for Langmuir adsorption isotherm.
11. Why is powdered charcoal is a better adsorbent than a lump of charcoal of the same mass.

QUESTIONS CARRYING FOUR MARKS

12. What are the differences between physical adsorption and chemical adsorption.
13. Explain how adsorption of gases takes place on to solids.
14. List the factors which affect the adsorption of gases on to solids
15. Describe the Langmuir's theory of adsorption of gases on to solid surface.
16. What is Freundlich's adsorption isotherm? How is it used to explain the adsorption of gases on solids.
17. Explain the use of adsorbent in catalysis.

18. How is an adsorbent used in gas masks? Explain the mechanism.
19. How is adsorption chromatography carried out using adsorbent material?
20. Write different kinds or adsorption isotherms.

10

BLOPOLYMERS

Polymers are very large molecules in size containing hundreds and thousands of monomers of organic origin. People have been using polymeric products since prehistoric time and chemists have been synthesizing them for the utility of society. Naturally occurring molecules are called Biopolymers which play a vital role in biological processes. Since his beginning man has been dependent on animal and vegetable matter for sustenance, shelter, warmth and other requirements and desires. Natural resins and gums have been used for thousands of years.

A polymer is a molecular compound distinguished by a high molecular mass, ranging into thousands and millions of Daltons and made up of repeating unit called a monomer. The physical properties of these macro molecules differ greatly from those of small molecules of organic origin. The special techniques have been used for studying the biopolymers. The techniques like spectroscopic techniques, light scattering, osmometry and other chromatographic techniques have been used in studying Biopolymers.

Polymerisation

This is a process in which monomers are joined to form the chain of a polymer. Apart from biopolymers, man has been synthesizing very large number of artificial or synthetic polymers. For example Polythene, Polyvinyl chloride, Polystyrene etc.

Classification

Polymers are mainly classified into two types. They are (1) Natural polymers, (2) Synthetic polymers.

1) Natural polymers: Natural polymers or biopolymers are the macromolecules present in biological systems. The well known biopolymers are proteins, carbohydrates and nucleic acids.

a) Proteins: Proteins are complex, natural compounds, composed of large number of different amino acids joined together by an amide linkage known as peptide linkage. The amino acids contain carboxyl as well as amino group. Therefore, one amino acid combines with another with help of their amino and carboxyl groups. It can be represented as

$$H_2N-\underset{R_1}{CH}-COOH + H_2N-\underset{R_2}{CH}-COOH \xrightarrow{-H_2O} H_2N-\underset{R_1}{CH}-\underbrace{\boxed{CO-NH}}_{\text{Peptide bond}}-\underset{R_2}{CH}-COOH$$

Since the resulting molecules still has a free amino and carboxyl group, it may further react with other amino acids at either of the ends to give a higher molecular weight condensation product. It is estimated that the human body contains about 100,000 different kinds of proteins each of which has a specific physiological function. They are present in the cytoplasm as well as cell membranes of all cells. Animals have relatively more proteins than plants in which polysaccharides are more. Only plants can make amino acids from organic materials like nitrates, ammonium sulphate, carbon dioxide and water, while most of the animals derive them mainly from plants and some other animals.

Polysaccharides

Polysaccharides or Carbohydrates are defined as the optically active polyhydroxy aldehydes or ketones or substances which yield these on hydrolysis. Carbohydrates are widely distributed in animals and plants. They serve as source of energy and also as store of energy (e.g., starch and glycogen). Certain carbohydrates (e.g., cellulose) support plant tissues, while some others (e.g., chitin) form the major constituents of the shells of crabs and lobsters. Cellulosic materials such as cotton, lignin, jute, grass, wood etc supply clothes, paper, fuel, plastics, paints and explosives.

Polysaccharides are formed by the polymerization of monomeric unit called monosaccharide. For example, starch is formed by polymerization of Glucose.

$$n\,C_6H_{12}O_6 \xrightarrow{\text{Polymerization}} (C_6H_{12}O_6)_n + n\,H_2O$$

The bond connecting the sugar units is called a glycosidic bond.

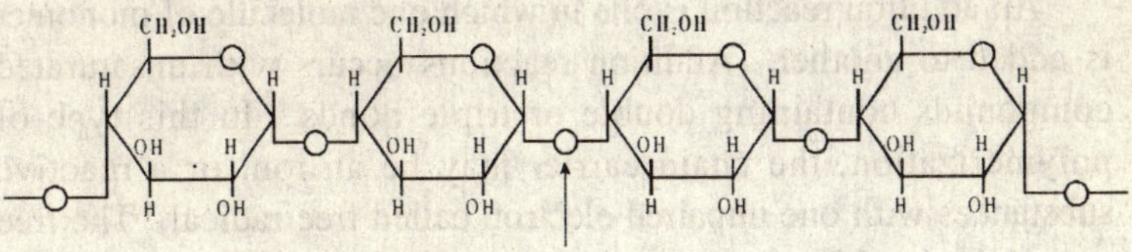

Fig 10.1 α - 1, 4 - Glycosidic bond in starch

Nucleic Acids

Nucleic acids are high molar mass polymers that play an essential role in protein bio synthesis. There are two types of nucleic acids present in the living organisms, they are Deoxy Ribonucleicacid (DNA) and Ribonucleicacid (RNA). DNA molecules are the largest biomolecules known. Its molar mass varies up to tens of billions of Daltons. RNA molecules are relatively smaller molecules.

DNA or RNA molecule contains only four types of building blocks, purines and pyrimidines, Furanose sugars and phosphate group. Each purine or pyrimidine are called a base. The repeating unit is called a nucleotide and hence nucleic acid is called a polynucleotide.

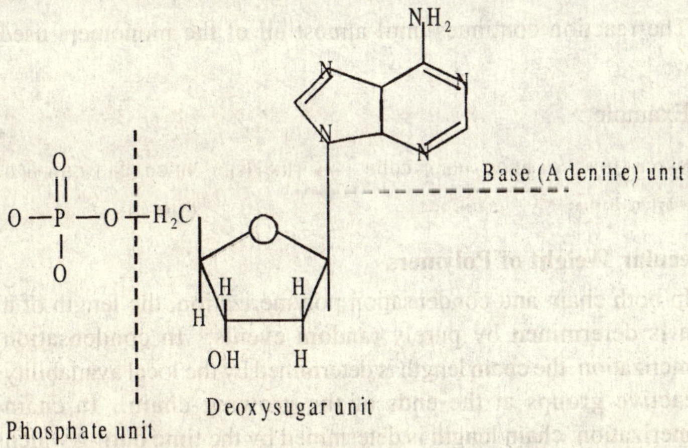

Fig 10.2: Structure of a nucleotide unit (Monomer unit)

Types of Polymerisation

The process of polymerization were divided into two types. They are addition and condensation polymerization.

1. Addition polymerization

An addition reaction is one in which one molecule of monomer is added to another. Addition reactions occurs with unsaturated compounds containing double or triple bonds. In this type of polymerization, the chain carrier may be an ion, or a reactive substances with one unpaired electron called free radical. The free radical is capable of reacting to open the double bond of a vinyl isomer and add to it, with an electron remaining unpaired. In a very short time, many more monomers added successively to the growing chain. Finally free radicals annihilate to form one or more polymer molecules.

Example : $n\,CH_2 = CH_2 \longrightarrow [CH_2 - CH_2]n$

Ethene — Polythene

2. Condesation Polymerization

In this type of polymerization, the condensation reaction takes place between two polyfunctional compounds to produce one larger polyfunctional molecule with the elimination of small molecule such as water, HCl, NH_3 etc.

The reaction continues until almost all of the monomers used up.

Example :

$$n\,H_2N-(CH_2)_6\,NH_2 + n\,HOOC-(CH_2)_4-COOH \longrightarrow [HN-(CH_2)_6-NH-CO-(CH_2)_4-CO] + n\,H_2O$$

Hexamethylene diamine — Adipic acid — Nylon 6.6

Molecular Weight of Polymers

In both chain and condensation polymerization, the length of a chain is determined by purely random events. In condensation polymerization, the chain length is determined by the local availability of reactive groups at the ends of the growing chain. In chain polymerization, chain length is determined by the time during which the chain grows before it diffuses into the vicinity of a second free radical and the two react.

In both cases of polymerization, the polymeric product contains molecules having many different chain lengths. The most important feature differentiating the polymers from low molecular weight species is the existence of a distribution of chain lengths and therefore, degrees of polymerization. Because of variation of chain lengths in a given polymer. Experimental measurement of molecular weight can give only average value. Due to these reasons, the molecular weights of polymers are calculated by two ways they are 1) weight average molecular weight and 2) Number average molecular weight.

1. Weight Average Molecular Weight ($\overline{M}_w$)

The experimental methods like light scattering experiments have given the result according to relative size of the polymer. In the averaging process, the molecular weight 'M' of each species is multiplied by the weight m of that species instead of number of molecules, therefore, the average molecular weight of a polymer is given by

$$\overline{M}_w = \frac{m_1 M_1 + m_2 M_2 + ... + m_i M_i}{m_1 + m_2 + ... m_i} = \frac{\Sigma mi\ Mi}{\Sigma mi}$$

Where m_1, m_2, etc are the weights of the polymer species with molecular weights m_1, m_2, etc.

Since weight of a species = Number of molecules of that species x its molecular weight i.e.,

$$mi = Nimi$$

$$\overline{M}_w = \frac{N_1 M_1^2 + N_2 M_2^2 + ... + Ni\,Mi}{N_1 M_1 + N_2 M_2 + ... Ni\,Mi} = \overline{M}_w = \frac{\Sigma\, Ni Mi^2}{\Sigma\, Ni Mi}$$

Where Ni is the number of polymer molecules of molar mass Mi in the sample.

Example: In a sample of a polymer, 50% of the molecules are monomers of molecular weight 40,000 and 50% are the monomers of molecular weight 60,000. Calculate weight average molecular weight.

Solution: weight average molecular weight $\overline{m}_w = \dfrac{\sum NiMi^2}{\sum NiMi}$

$$N_1 M_1^2 = \frac{50}{100} \times (40{,}000)^2$$

$$N_2 M_2^2 = \frac{50}{100} \times (60{,}000)^2$$

$$\therefore \overline{M}_w = \frac{\frac{50}{100} \times (40{,}000)^2 + \frac{50}{100} \times (60{,}000)^2}{\frac{50}{100} \times 40{,}000 + \frac{50}{100} \times 60{,}000}$$

$$\overline{M}_w = 52{,}000$$

Number Average Molecular Weight

Generally the polymer molecules does not contain molecules of same chain lengths. They have distribution of chain lengths and hence, they possess a range of molecular weights. The molecular weight determined on the basis of colligative properties suggest that they depend on the number of polymer molecules in solution. Such average molecular weight can be measured by the osmotic pressure determination method, which is mentioned at the end of this chapter.

The number average molecular weight can be expressed by a mathematical formula

$$\overline{M}_n = \frac{n_1 M_1 + n_2 M_2 + \dots\dots\dots + ni\,Mi}{n_1 + n_2 + \dots\dots\dots + ni}$$

Where n_1, n_2 etc are the numbers of molecules with molecular weights M_1, M_2 etc.

Thus the above expression can also be expressed as

$$\overline{M}_n = \frac{\Sigma NiMi}{\Sigma Ni}$$ Where Ni is the number of polymer molecules with molecular weights mi in the sample.

Example - 2: A polymer sample consists of 30% of molecules with 70,000 molecular weight and 70% of the molecules with 60,000 molecular weight. Calculate the number average molecular weight?

$$\overline{M}_n = \frac{\Sigma NiMi}{\Sigma Ni} = \frac{N_1 M_1 + N_2 M_2}{N_1 + N_2}$$

$N_1 = 30 \quad M_1 = 70{,}000$

$N_2 = 70 \quad M_2 = 40{,}000$

$$\overline{M}_n = \frac{30 \times 70{,}000 + 70 \times 66{,}000}{30 + 70}$$

$$\overline{M}_n = 67{,}200 \text{ g } \overline{\text{mol}}^1$$

Example – 3: In a polymer sample 30% molecules have molecular mass 20,000, 40% have molecular mass 30,000 and the rest 30% have 60,000. Calculate the number average and weight average molecule or weight.

$N_1 = 30 \quad M_1 = 20{,}000$

$N_2 = 40 \quad M_2 = 30{,}000$

$N_3 = 30 \quad M_3 = 60{,}000$

$$\overline{M}_n = \frac{N_1 M_1 + N_2 M_2 + N_3 M_3}{N_1 + N_2 + N_3}$$

$$\overline{M}_n = \frac{(30 \times 20{,}000) + (40 \times 30{,}000) + (30 \times 60{,}000)}{30 + 40 + 30}$$

$$\overline{M}_n = 36{,}000$$

$$\overline{M}_n = \frac{30(20{,}000)^2 + 40(30{,}000)^2 + 30(60{,}000)^2}{(30 \times 20{,}000) + (40 \times 30{,}000) + (30 \times 60{,}000)}$$

$$\overline{M}_n = 43{,}333$$

Natural rubber

It is a natural polymer made up of monomer known as isoprene. Chemically rubber is 1,4 polymer of isoprene. It can be represented as.

$$n\,CH_2 = \underset{\textstyle\text{Isoprene}}{\overset{\overset{\displaystyle CH_3}{|}}{C}} - CH = CH_2 \xrightarrow{\text{Polymerisation}} \left[CH_2 - \overset{\overset{\displaystyle CH_3}{|}}{C} = CH - CH_2 \right]_n$$

Isoprene — Poly isoprene

It has remarkable elasticity and undergoes long range reversible extension even under relatively small application of force. It is produced by the latex of the rubber plant. The latex contain rubber particles in the form of a colloidal solution. This plant is grown in tropical and semitropical countries like, India, Srilanka, Malasia, Indonesia etc.

Natural rubber is used to make souls for chappals, tyres etc.

Determination of Molecular Weight by Light Scattering

Modern light scattering instruments use either a mercury arc or a laser as a source and detect the scattered light photoelectrically. Many are interfaced to computer for central data handling and computation of results. The essential features of a light scattering photometer widely used Brice phoenix instrument (fig. 10.3). Light from a source (s) passess through a lense (L), a polarizer (P_1) and a monochromatic filter (F). It is then strikes either a calibrated reference standard or a glass cell (c) holding the polymer solution. After passing through the cell, the primary light beam is absorbed in a light trap tube (T). Scattered light from cell or standard is viewed by a multiplier phototube (R) after passing through a slit system D_1 - D_2 and polarizer (P_2). The photo tube is powered by high voltage from a regulated electronic power supply. Its out put signal is transmitted either to a strip - chart recorder or to an analogue to digital converter for computer use.

A scattering, cell consists of a glass cylinder having flat entrance and exist windows. The scattering cell is centered on the axis of rotation of the receiver photo tube.

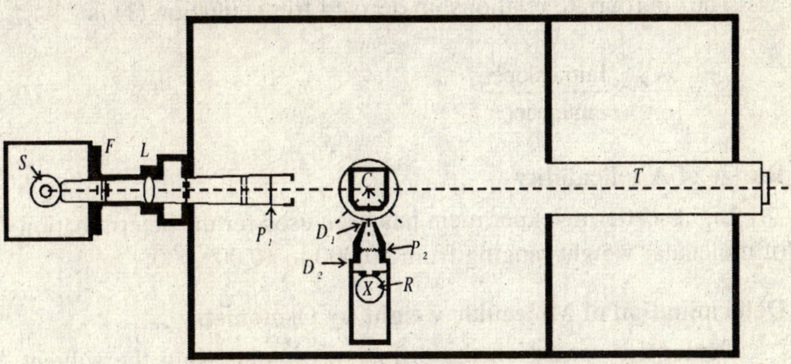

Fig 10.3: Light Scattering Photometer

Sample Preparation

The solvents and polymer solutions must be clarified by filtration or by ultracentrifugation, in order to remove dust particles and other extraneous particles. Accuracy in light scattering and measuring is favoured by the proper choice of the solvent. The difference in the refractive index between polymer and solvent should be as large as possible.

Calibration: In principle observed scattered intensity can be related to turbidity through knowledge of the geometry of the photometer. Calibration with substances of known turbidity is a common practice. These substances include reflecting standards colloidal substances, and simple liquids tugstosilicic acid

($H_4SiW_{12}O_{40}$, M = 2879),

Treatment of Data

The weight - average molecular weight is the inverse of the intercept at c = 0 and θ = 0 in the zimmplot the second virial coefficient is calculated from the slope of the lines at constant angles by equation.

$$K\frac{C}{\Delta R_{90}} = H\frac{C}{\Delta T} = \frac{1}{M_w P_{(\theta)}} + 2A_2C + \ldots\ldots\ldots\ldots\ldots\ldots(1)$$

The radius of gyration is derived from the slope of the zero concentration line as a function of angle.

The appropriate relationship derived from equation (1) is

$$S_Z^2 = \frac{3\lambda^2 s}{16\pi^2} \times \frac{\text{Initial slope}}{\text{Intercept}}$$

Range of Applicability

Light scattering experiment has been used for the determination of molecular weight ranging from 10,000 to 10,000,000.

Determination of Molecular weight by Osmometry

Osmometry based on the diffusion of solute into the solvent through semi permeable membrane. The diffusion of solute into solvent is a bilateral process. It consists of (1) the solute molecule moving up into solvent and (2) The solvent molecules moving down into solution. This intermingling of solute and solvent molecules goes in until the equilibrium is established. Thus diffusion of solute will take place when two solutions of unequal concentration are in contact. The rate of diffusion is directly dependent on the molecular weight of the solute. The flow of the solvent through a semi permeable membrane from pure solvent to solution or from a dilute solution to concentrated solution is known as osmosis. The osmotic pressure of a solution can be determined experimentally by using the apparatus called osmometers.

Principle: The osmotic pressure measurements are useful in determining the molecular weight of compounds like polymers which have high molecular weights. Knowing the osmotic pressure of a given solution, the molecular weight of the solute can be calculated as follows from vant Hoff equation.

$$\pi = \text{Nrt}$$

$$\pi V = \frac{W}{M} RT$$

$$M = \frac{WRT}{\pi V}$$

Where M = Molecular mass of solute.
W = Amount of solute in grams.
R = Gas constant.
T = (t^0c + 273)k
π = Osmotic pressure in Atmospheres.
V = Volume of solution in Liter.

A modern apparatus for the determination of osmotic pressure is represented in the diagram 10.4. It consists of a stainless steel container in which a semi permeable membrane is fixed. The membrane is partition between solute and solvent compartments. A diaphragm which is flexible to changes in osmotic pressure is fixed in the solvent compartment. When the solution containing polymer and the solvent in which the polymer is soluble is taken in two separate compartments and the taps are closed. Osmosis occurs. The solvent flows into solution across the semi permeable membrane. This reduces the pressure in the solvent compartment causing the diaphragm to distort eventually, the pressure becomes low enough to stop occurrence of osmosis. The degree of distortion is related to the osmotic pressure of the solution. The extent of distortion is measured by a device known as strain gauge. The strain gauge provides an electric current that is proportional to the extent of distortion. The gauge is calibrated to give osmotic pressure directly.

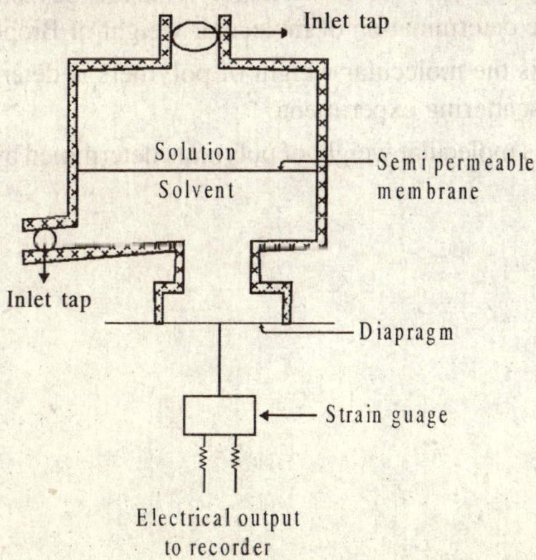

Fig. 10.4 : Osmometer

QUESTIONS CARRYING TWO MARKS

1. What is a polymer? Give one example?
2. Give two examples of polymers.
3. What are Biopolymers? Give examples.
4. How are polymers classified?
5. What is addition polymerization? Explain with an example.
6. What is condensation polymerization? Explain with an example.
7. Write a brief note on proteins as Biopolymers.
8. Write a brief note on carbohydrates as polymers
9. Write a brief note on Nucleic acids as Biopolymers.
10. Discuss the constitution of rubber.
11. What is weight average molecular weight? Write the mathematical expression.
12. What is number average molecular weight? Write the expression for calculating the number average molecular weight?

QUESTIONS CARRYING FOUR MARKS

13. Name four types of Biopolymers. What are the methods used for the determination of molecular weight of Biopolymers.
14. How is the molecular weight of polymers is determined by light scattering experiment.
15. How is molecular weight of polymers determined by osmometry.

11

VISCOSITY

Defination: The resistance which a liquid exerts against the displacement of its own molecules is known as viscosity. It is well known that all liquids do not flow with the same speed. Some liquids like water, ether etc. flow rapidly while some other liquids like glycerin, groundnut oil etc. flow very slowly. This indicates that every liquid has some internal resistance to flow.

The liquids which flow relatively slowly will have high internal resistance and therefore, are said to be more viscous or have high viscosity. On the other hand the liquids which flow rapidly have low internal resistance and hence are said to be less viscous i.e., their viscosity is less.

To understand the nature of the internal resistance or friction within a liquid, consider a liquid flowing through a narrow tube. All parts of the liquid do not move through the tube in the same speed. Imagine the liquid made up of cylindrical coaxial layers. The layer which is in contact with waves of the tube is almost stationary.

As we move from the walls towards the centre of the tube the viscosity of the cylindrical layer keeps on increasing till it is maximum at the centre. This means as we move from the centre towards the walls, the velocity of the layers keep on decreasing. As a consequence of this, every layer offers some resistance or friction to the layer immediately below it. This force of friction which one layer of the liquid offer to another layer moving with a different velocity is called viscosity.

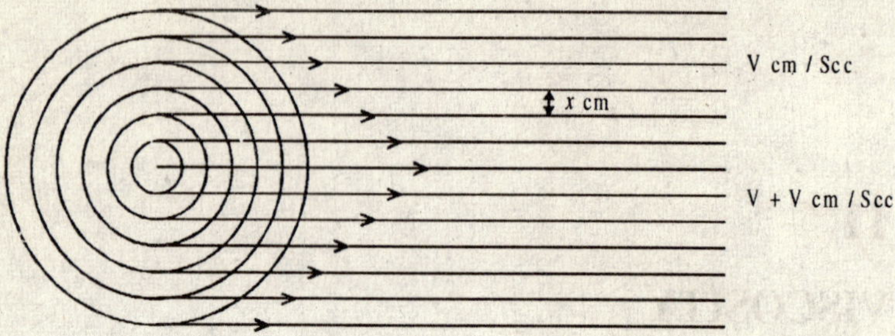

Fig. 11.1: Flow of liquid through a narrow tube

It has been found that force of friction (f) between two cylindrical layers each having the area 'A' sq.cm. separated by a distance x cm and having a velocity difference v cm/sec is given by

$$f \propto A\, V/x \text{ or } f = \eta \frac{AV}{x}$$

Where η is a constant known as coefficient of viscosity.

If x = 1cm A = 1 sq.cm. and V = 1cm/ sec then $f = \eta$

therefore, the coefficient of viscosity is defined as the force of friction (in dynes) required to maintain the velocity difference of 1 cm / sec between two parallel layers, 1 cm apart and each having an area of 1 sq.cm.

Factors Affecting Viscosity

The factors affecting the viscosity of a liquid are

a) Temperature
b) The presence of solutes, lyophilic colloids and other suspended impurities tend to increase the viscosity of the liquids
c) Polarity of compounds also affects the viscosity.
d) Branched chain compounds possess greater viscosity than the straight chain ones.
e) Increase in molecular weight increases the viscosity of the

liquid

f) P^H of the medium in which compounds are placed.

g) Pressure of the liquid.

Measurement of Viscosity

The viscosity of the liquid is usually measured by observing the rate of flow of liquid through some type of capillary tube. The rate of flow of liquid must remain steady and should be parallel to the axis of the tube. Besides these, the rate of flow should not exceed certain value which depends up on the radius of the tube as well as the viscosity of the liquid. Poiseuille derived a law for determining the coefficient of viscosity and is based on the viscous flow of liquids through capillary tubes. According to poiseuille,

$$\eta = \frac{\pi P r^4 t}{8 v l}$$

Where P = pressure.

r = Radius of the tube in cm.

t = Time of flow of the liquid

v = Volume of the liquid flowing across the whole cross section of the capillary tube.

l = Length of the tube in cm.

η = Coefficient of viscosity.

The value of absolute viscosity can be determined directly by means of poiseuille's expression. After knowing the rate of flow of liquid through a capillary tube of uniform size and known dimensions.

Since this direct method is difficult to perform, generally relative viscosity is determined by taking a reference standard. i.e., the liquid of known viscosity.

Determination of Viscosity of a Liquid by Ostwald's Viscometer

Principle Viscosity is the resistance of a liquid to flow. Higher the viscosity, the longer the time taken by certain volume of the liquid to flow. Using poiseuilles equiation for viscosity we can relate the viscosity of the liquid to viscosity of water to find the relative viscosity of a liquid, we need to obtain the time of flow of a certain volume of that liquid and the time of flow of the same volume of water. This can be done using ostwald's viscometer. The density of the liquid is also determined. The viscosity of water and the density of water is obtained from the published data.

$$\text{Viscosity of liquid } (\eta_L) = \frac{D_L}{D_w} \times \frac{T_L}{T_w} \times \eta_w$$

D_L = Density of liquid

D_W = Density of water

T_L = Time of flow of liquid

T_W = Time of flow of water

η_W = Viscosity of water.

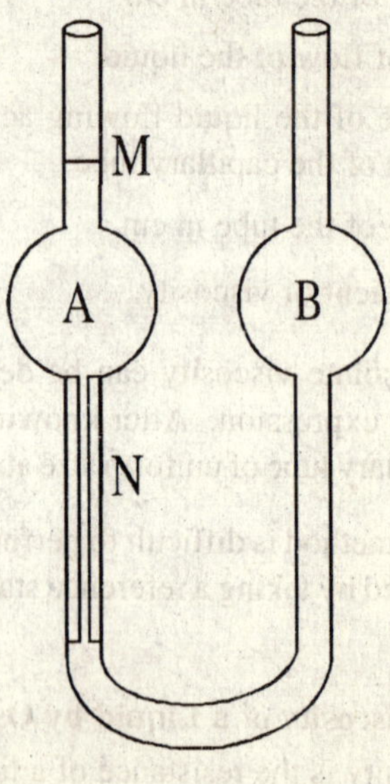

Fig. 11.2 : Ostwald's Viscometer

Procedure: pipette out a known volume of the liquid into a clean, dry viscometer that has been clamped vertically in a constant temperature both so that the mark etched on the viscometer are visible clearly. A small rubber tube is attached to the narrow limb. The viscometer along with liquid is allowed to attain the water bath temperature. The liquid is then sucked into the bulb well above the mark, with the help of rubber tube. It is allowed to flow through the capillary tube and the time of flow from the upper to the lower mark is noted using a stop watch the experiment is repeated.

The liquid in the viscometer is taken out and clean the viscometer and repeat the same experiment for water also.

Determination of Density of a Liquid

Density of a liquid is the mass per unit volume. When we use the term density at a given temperature, it means the relative density at that temperature with respect to the density of water. The density of water is taken as unity in all practical determinations.

The density of liquid can be measured by using Pyknometer or specific gravity bottle.

The Pyknometer or specific gravity bottle is washed with chromic acid solution and then with distilled water and finally with alcohol. It is then dried. The apparatus is weighed in an analytical balance. Then it is filled with the liquid and weighed again. The difference in weight will give weight of the liquid. Similarly the weight of water is also recorded. The density of the liquid is determined by following equation.

$$\frac{d_1}{d_2} = \frac{w_1}{w_2} \qquad d_1 = \frac{W_1}{W_2} \times d_2$$

d_1 = Density of liquid

W_1 = Weight of liquid

d_2 = Density of water

W_2 = Weight of water

Relation of Viscosity and Shape of Molecules

The viscosity is affected by size and shape of the molecules present in a given solution. The compounds with high molecular weight will have higher viscosity. Similarly the compound with branched chain will have lower viscosity than the linear chain. One of the important application of viscosity measurements is in determination of molecular weight of polymers.

The polymeric compound can be dissolved in a suitable solvents and the viscosity of the solution is determined using ostwald's viscometer. The molecular weight of high polymer can be calculated according to mark - Houwink equation.

$[\eta]=KM^x$ Where M is the molecular weight of the polymer and K is a constant for a polymer- solvent system at a definite temperature.

x is also a constant known as shape factor.

The value $[\eta]$ is known as intrinsic viscosity which is obtained as intercept of the curve

between $\frac{nsp}{C}$ and C,

Where c = concentration of polymer, calculated by means of equation

$$\eta_{sp} = \frac{\eta solution - \eta solvent}{\eta solvent.}$$

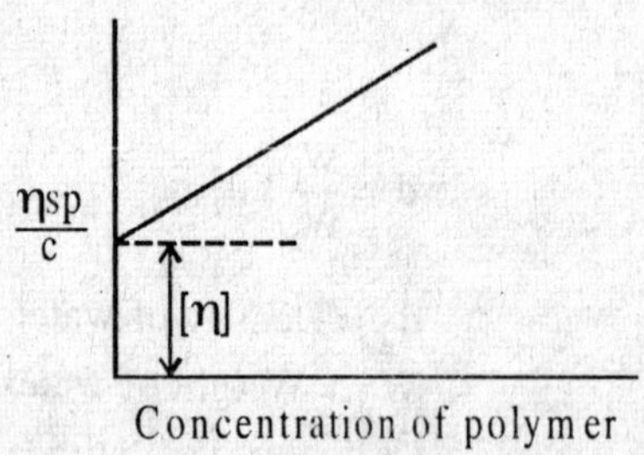

Fig. 11.3 : Viscosity of Polymers

Coefficient of Viscosities for some common liquids

Liquid	*Viscosity*
Ethyl ether	2.33 millipoises
Acetone	3.29 millipoises
Methanol	5.93 millipoises
Chlorobenzene	8.0 millipoises
CCl_4	9.68 millipoises
Ethanol	12.0 millipoises
Nitrobenzene	20.3 millipoises

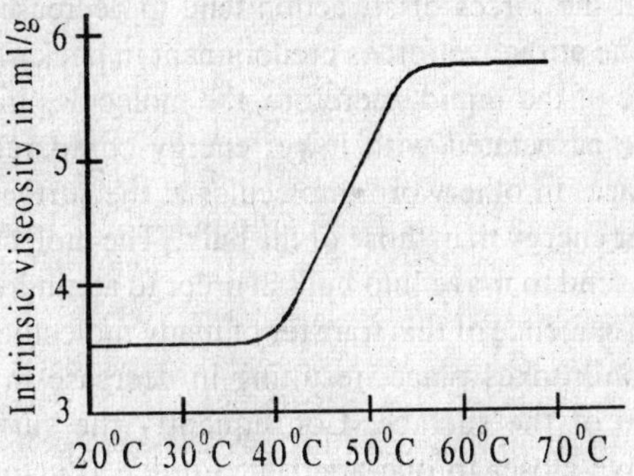

QUESTIONS

1. Define the term viscosity of a liquid
2. What are the factors affects viscosity of a liquid?
3. How is viscosity of a liquid is determined by using oswald's viscometer
4. Explain the terms viscosity coefficient of a liquid? Describe the method for the determination of it.
5. How is density of a liquid is determined? How is it related to viscosity of a liquid?
6. Describe how is viscosity related to shape of the molecules.

12

SURFACE TENSION

The surface tension of a liquid is defined as the force acting at right angle to the surface along one centimeter length of the surface. Thus the units of surface tension is dynes per centimeter or Newtons per metre (NM^{-1}).

It is observed that the forces of attraction tend to decrease the energy of a system. The attractive forces predominant in buck of the liquid than the surface of the liquid, therefore, the molecules in the bulk of the liquid are associated with lower energy compared to molecules of the surface. In otherword, molecules at the surface of a liquid possess greater energy than those of the bulk. The molecules present on the surface tend to move into bulk in order to attain lower energy state. As a consequence of this transfer of many molecules of the surface into the bulk takes place, resulting in decrease in the number of molecules of the surface. Consequently the surface molecules tend to move closer to one another in order to gain the normal distance between them. This process results in contraction of a surface of liquid. Therefore, the surface of a liquid behaves as if it were in a state of tension. The force that operates on the surface of a liquid to contract the liquid is known as surface tension.

The process of surface tension can be explained by reference fig. 12.1

Fig. 12.1: Surface tension of a liquid

Consider a molecule 'R' which is present in the bulk of the liquid. This is attracted equally in all the directions by other molecules. Therefore the net force acting on it is not there. Consider the next molecule 'P' which is on the surface of the liquid. The down ward force of attraction by the molecules of the liquid is more than upward forces since the number of molecules below is more than in the air above the surface. These unbalanced attractive forces acting downward tend to draw the surface molecules into the body of the liquid and therefore surface tension is produced.

Factors Affecting the Surface Tension

1. Temperature

When the temperature of the liquid is raised, the surface tension decreases.

2. Effect of solutes on surface tension of solvent

Some substances on addition to a solvent known to decrease the surface tension, they are called **capillary active**. e.g., soaps, proteins, organic acids, alcohols etc.

Some substance are known to increase the surface tension of a solvent they are known as **capillary inactive** substances. e.g., inorganic salts, bases, sugars, glycerine etc.

The capillary activity of amphipathic molecules e.g., soaps largely dependent on the nature of nonpolar chain. They tend to accumulate on the surface of the liquid leading to decrease in surface tension.

Inorganic electrolytes, raise the surface tension of water as they are negatively adsorbed on the surface of water.

Measurement of Surface Tension

Surface tension of liquids can be measured by two types of methods 1) static and 2) Dynamic. The static method is based on the assumption that the liquid has attained surface equilibrium. For pure liquids and solutions the static method is more suited. Where as for colloidal solutions dynamic methods are more suited.

Among the static methods, the most commonly used are

1) The capillary rise method

2) The duNouy rising method
3) The withelmy balance method.
4) The drop weight method

In the present syllabus the stalagmometer method is explained for the determination of surface tension.

Determination of surface tension of a liquid by using stalagmometer

Principle: Surface tension is the property of a liquid that causes the liquid to have the least surface area. The drop weight method is based on the principle that the weight of a drop of the liquid falling from a capillary tube held vertically is approximately proportional to the surface tension of the liquid. The higher the surface tension, the less the drops that will be formed from a certain volume of the liquid. To find the relative surface tension of a liquid, we need to obtain the number of drops formed from the same volume of water. This can be done using stalagmometer. The density of liquid is also determined by using specific gravity bottle. The surface tension of water and density is obtained using the published data.

Procedure : A Clean, dry, Stalagmometer is clamped vertically with the flattened end dipping into a beaker containing the given liquid a rubber tubing having a screw clip is attached to the upper end. The liquid is carefully sucked into the stalagmometer till it is above the upper mark and the screw clip is adjusted to allow a drop rate 15 to 20 drops per minute. The liquid is sucked again and allowed to flow and the number of drops delivered from the time the meniscus leaves the upper mark till it reaches the lower mark is counted. This is repeated. The stalagmometer is washed with acetone and then with water, dried. The number of drops formed for water also determined in the same way and the surface tension is calculated using the following equation.

Surface tension of liquid $$\gamma_L = \frac{D_L}{D_w} \times \frac{N_w}{N_L} \times \gamma_w$$

Where D_L = Density of liquid

D_w = Density of water

N_w = Number of drops of water formed

γ_w = Surface tension of a liquid.

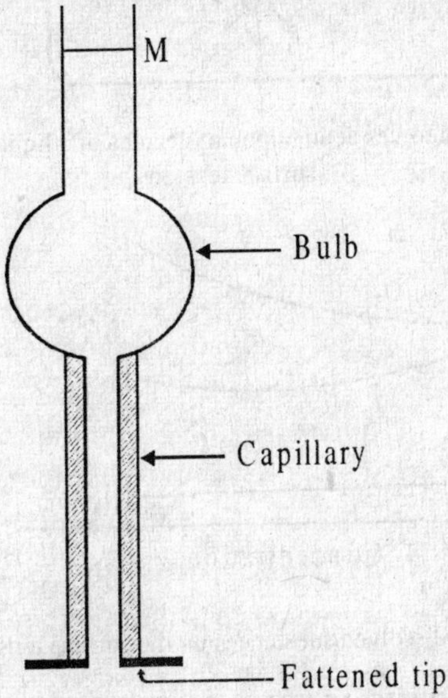

Fig 12.2: Stalagmometer

Effect of Surfactants on Surface Tension

There are certain substances, which are known to lower the surface tension when they are added to a solvent, they are known a surface active reagents or surfactants. Compounds like soaps, sulphonic acids, methyl alcohol, ethyl alcohol acetone etc, when added to water, lower its surface tension. The lowering of surface tension by the addition of above substances can be explained as follows. If a greasy substances sticking on the surface of the cloth, water alone cannot remove such materials because both are not miscible. However, when soap is added, the interfacial tension between water and soap is reduced and the mixing of the two takes place. In other wards emulsification of grease in water takes place. Then the mechanical action like rubbing, releases the dirt. Surface active substances are also added to medicinal emulsions, toothpastes, mouth washes and toilet creams in order to enable them to spread evenly there by increasing the efficiency of that product.

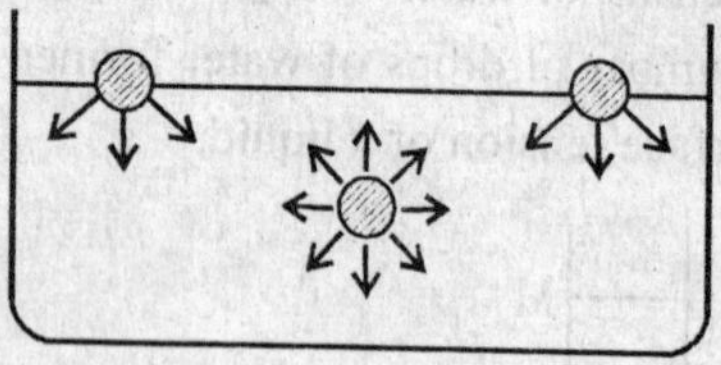

Fig. 12.3: Cohesive forces acting upon molecules of a liquid so as to result in surface tension

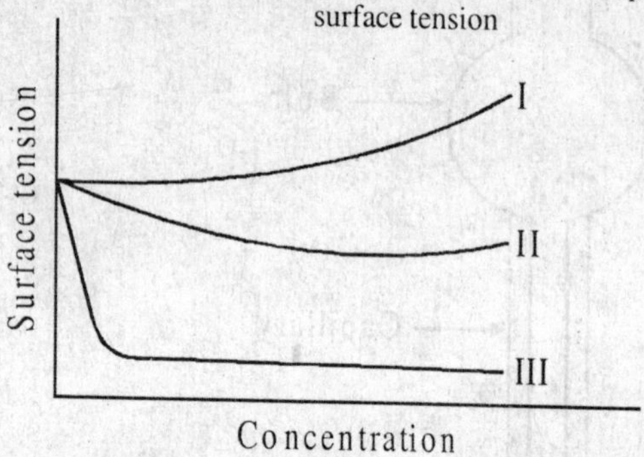

Fig. 12.4: Effect of dissolved substances on the surface tension of the solvent

TWO MARKS QUESTIONS

1. Define the surface tension of a liquid.
2. What are surfactants? Give one example.
3. Name the factors which affect surface tension of a liquid.
4. Name any two methods used for the determination of surface tension.
5. What are the factors affecting the surface tension of a liquid?

FOUR MARK QUESTIONS

6. Describe the cause for the formation of a liquid drop.
7. How do you determine the surface tension of a liquid using stalagmometer?
8. How does the surfactants bring down the surface tension of a liquid.

APPENDIX

BANGLORE UNIVERSITY

B.Sc. Biochemistry Syllabus

1 SEMESTER

Paper 1.1; Biochemistry - 1

1. Measurement 4 h

International system of units - derived units, subsidiary units. Significant figures. Dimensional analysis. exponential notation. Graphical representation of data. Errors in quantitative analysis.

2. Atomic structure 4 h

Electromagnetic radiation, Wave particle duality - the de Broglie equation. The Heisenberg uncertainty principle. The schrodinger wave equation. Quantum numbers. Atomic orbitals and their shapes. The Pauli exclusion principle. Hund's rule. The Aufbau process. Electronic configuration of the elements.

3. Periodic Classification 3 h

Trends in the periodic table. Atomic and ionic radii, ionization energy, electro negativity and electron affinity. Concept of oxidation number and its computation.

4. Chemical Bonding 8 h

Ionic bond - energetics, Born Haber cycle. Covalent bond - Valence bond theory. Hybridization - examples: methane, ammonia, water, ethane and ethylene. Sigma and Pi bonds. Concept of resonance. Molecular orbital theory. Properties of covalent molecules, bond length. Bond energy and bond angle. Polarity of molecules. Coordinate bond. Vander waal's forces. Hydrogen bonds; inter-and intra molecular types, importance in biomolecules. Hydrophobic forces.

5. Radioactivity 7 h

Natural and artificial radioactivity. Characteristics of radio elements. Disintegration constant. Half-life. Tracer techniques -production of labelled ^{14}C, ^{32}P and compounds and their applications in biological studies. Biological effects of radiation emitted by isotopes, safety measures.

6. Solutions and Colligative Properties 6 h

Concentration units - molarity, molality, normality and mole fraction. Types of solutions, - homogenous and heterogenous. Factors influencing solubility, solubility curves. Henry Law - applications. Osmotic pressure and its measurement by the Berkley - Hartley method. Laws of osmotic pressure. Hypo-, hyper and isotonic solutions. effect of osmotic pressure on living cells. Donnan membrane equilibrium. Relative lowering of vapour pressure, Rault's law. Elevation of boiling point, depression in freezing point, and their applications in the determination of molecular weight. Abnormal molecular weights. Vant Hoff factor. Degree of association and Degree of dissociation.

7. Electrolytic Dissociation and Mass Law 8 h

Strong and weak electrolytes. Activity and activity coefficient. Relationship between activity coefficient and ionic strength. Common ion effect, solubility product and their applications. Conductance and its measurement. Electrochemical cells Oxidation -reduction reactions, reversible electrodes and cells. Single electrode potential. Nernst equation. Standard electrode potentials. Electrochemical series and applications. reference electrodes. Ion selective electrodes and their applications.

8. Acids, Bases and Buffers 6 h

Modern concepts of acids and bases. Ionization of acids. Dissociation of water. Ionic product of water. Hydrogen ion concentration - pH. Determination of pH. Dissociation of weak acids. Effects of salts on dissociation of acids. Interaction of acids with bases.

Dipolar ions and isoelectric pH of amino acids and proteins. Buffer equation, buffer capacity. Problems on preparation of buffer solutions. Buffers of blood plasma, red blood cells and tissue fluids. Use of buffers. Determination of pH - colorimetric method based on use of indicators, limitations. Use of buffer solutions and indicators.

9. Adsorption 3 h

Adsorption of gases by solids. Heat of adsorption

Freundlich and Langmuir adsorption isotherms with derivations. Applications of adsorption.

10. Biopolymers 3 h

Classification. The polymerization process. Number -average and weight - average molecular weights. Molecular weight determination by osmometry and light scattering techniques.

11. Viscosity of liquids 2 h

Determination of viscosity of liquids using Oswald's viscometer. Relation of viscosity and shape of molecules, examples.

12. Surface tension 1h

Definitions, determination of surface tension of liquids using stalagmometer. Effect of surfactants.